MENU-M

D1538837

presents

The Old-Fashioned Ice Cream Parlor

Book 2
Multiplication and Division

A Remedia Publication

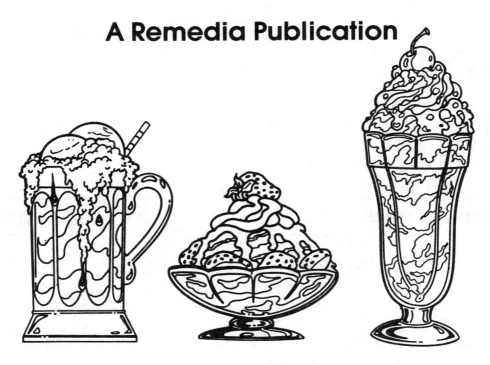

Created by
Barbara Johnson
Kitty Scharf

Illustrations by
Edwin Vandermulen

REMEDIA PUBLICATIONS   **10135 E. VIA LINDA #D124**   **SCOTTSDALE, AZ 85258**

To the teacher: For additional practice, blank checks are provided so students may create their own checks using the menu and the tax table on page 24.

Ice Cream Parlor

	Subtotal	
	Tax	
	Total	

Ice Cream Parlor

	Subtotal	
	Tax	
	Total	

Ice Cream Parlor

	Subtotal	
	Tax	
	Total	

Ice Cream Parlor

	Subtotal	
	Tax	
	Total	

Name _____

**Multiplying 2-digit
numbers by 1 digit
(2's, 3's, 4's, 5's)**

Example:

```
          1
  $  .26
  ×    2
  $  .52
```

1.

```
  $  .22          $  .26          $  .33          $  .21
  ×    2          ×    3          ×    2          ×    5
```

2.

```
  $  .21          $  .32          $  .12          $  .13
  ×    4          ×    2          ×    3          ×    4
```

3.

```
  $  .25          $  .35          $  .14          $  .16
  ×    2          ×    3          ×    4          ×    5
```

4.

```
  $  .37          $  .29          $  .32          $  .18
  ×    4          ×    3          ×    5          ×    4
```

5.

```
  $  .27          $  .25          $  .41          $  .43
  ×    4          ×    4          ×    2          ×    4
```

6.

```
  $  .18          $  .16          $  .28          $  .21
  ×    5          ×    2          ×    3          ×    3
```

MENU-MATH/ICE CREAM PARLOR

**Multiplying 2-digit
numbers by 1 digit
(2's, 3's, 4's, 5's)**

1.

$\begin{array}{r} \$\ .75 \\ \times\ 2 \\ \hline \end{array}$ $\begin{array}{r} \$\ .48 \\ \times\ 3 \\ \hline \end{array}$ $\begin{array}{r} \$\ .62 \\ \times\ 4 \\ \hline \end{array}$ $\begin{array}{r} \$\ .61 \\ \times\ 3 \\ \hline \end{array}$

2.

$\begin{array}{r} \$\ .36 \\ \times\ 2 \\ \hline \end{array}$ $\begin{array}{r} \$\ .46 \\ \times\ 4 \\ \hline \end{array}$ $\begin{array}{r} \$\ .75 \\ \times\ 4 \\ \hline \end{array}$ $\begin{array}{r} \$\ .75 \\ \times\ 3 \\ \hline \end{array}$

3.

$\begin{array}{r} \$\ .37 \\ \times\ 4 \\ \hline \end{array}$ $\begin{array}{r} \$\ .66 \\ \times\ 2 \\ \hline \end{array}$ $\begin{array}{r} \$\ .85 \\ \times\ 3 \\ \hline \end{array}$ $\begin{array}{r} \$\ .92 \\ \times\ 2 \\ \hline \end{array}$

4.

$\begin{array}{r} \$\ .81 \\ \times\ 3 \\ \hline \end{array}$ $\begin{array}{r} \$\ .52 \\ \times\ 4 \\ \hline \end{array}$ $\begin{array}{r} \$\ .86 \\ \times\ 4 \\ \hline \end{array}$ $\begin{array}{r} \$\ .44 \\ \times\ 5 \\ \hline \end{array}$

5.

$\begin{array}{r} \$\ .94 \\ \times\ 4 \\ \hline \end{array}$ $\begin{array}{r} \$\ .45 \\ \times\ 3 \\ \hline \end{array}$ $\begin{array}{r} \$\ .83 \\ \times\ 2 \\ \hline \end{array}$ $\begin{array}{r} \$\ .81 \\ \times\ 5 \\ \hline \end{array}$

6.

$\begin{array}{r} \$\ .62 \\ \times\ 3 \\ \hline \end{array}$ $\begin{array}{r} \$\ .65 \\ \times\ 2 \\ \hline \end{array}$ $\begin{array}{r} \$\ .53 \\ \times\ 4 \\ \hline \end{array}$ $\begin{array}{r} \$\ .61 \\ \times\ 5 \\ \hline \end{array}$

Name _____

Multiplying 2-digit numbers by 1 digit (8's, 9's)

Example:

$$\begin{array}{r} 7 \\ \$\ .28 \\ \times\quad 9 \\ \hline \$\ 2.52 \end{array}$$

1.

$$\begin{array}{r} \$\ .21 \\ \times\quad 8 \\ \hline \end{array}$$
$$\begin{array}{r} \$\ .23 \\ \times\quad 9 \\ \hline \end{array}$$
$$\begin{array}{r} \$\ .42 \\ \times\quad 8 \\ \hline \end{array}$$
$$\begin{array}{r} \$\ .49 \\ \times\quad 9 \\ \hline \end{array}$$

2.

$$\begin{array}{r} \$\ .25 \\ \times\quad 8 \\ \hline \end{array}$$
$$\begin{array}{r} \$\ .26 \\ \times\quad 9 \\ \hline \end{array}$$
$$\begin{array}{r} \$\ .15 \\ \times\quad 8 \\ \hline \end{array}$$
$$\begin{array}{r} \$\ .28 \\ \times\quad 9 \\ \hline \end{array}$$

3.

$$\begin{array}{r} \$\ .35 \\ \times\quad 9 \\ \hline \end{array}$$
$$\begin{array}{r} \$\ .31 \\ \times\quad 8 \\ \hline \end{array}$$
$$\begin{array}{r} \$\ .38 \\ \times\quad 8 \\ \hline \end{array}$$
$$\begin{array}{r} \$\ .41 \\ \times\quad 9 \\ \hline \end{array}$$

4.

$$\begin{array}{r} \$\ .36 \\ \times\quad 8 \\ \hline \end{array}$$
$$\begin{array}{r} \$\ .27 \\ \times\quad 8 \\ \hline \end{array}$$
$$\begin{array}{r} \$\ .47 \\ \times\quad 8 \\ \hline \end{array}$$
$$\begin{array}{r} \$\ .43 \\ \times\quad 9 \\ \hline \end{array}$$

5.

$$\begin{array}{r} \$\ .16 \\ \times\quad 9 \\ \hline \end{array}$$
$$\begin{array}{r} \$\ .14 \\ \times\quad 8 \\ \hline \end{array}$$
$$\begin{array}{r} \$\ .52 \\ \times\quad 8 \\ \hline \end{array}$$
$$\begin{array}{r} \$\ .50 \\ \times\quad 9 \\ \hline \end{array}$$

6.

$$\begin{array}{r} \$\ .36 \\ \times\quad 9 \\ \hline \end{array}$$
$$\begin{array}{r} \$\ .43 \\ \times\quad 8 \\ \hline \end{array}$$
$$\begin{array}{r} \$\ .63 \\ \times\quad 8 \\ \hline \end{array}$$
$$\begin{array}{r} \$\ .34 \\ \times\quad 9 \\ \hline \end{array}$$

Name _____

Multiplying 2-digit numbers by 1 digit (8's, 9's)

1.
$$\begin{array}{r} \$\ .66 \\ \times\ \ 9 \\ \hline \end{array}$$
$$\begin{array}{r} \$\ .55 \\ \times\ \ 8 \\ \hline \end{array}$$
$$\begin{array}{r} \$\ .65 \\ \times\ \ 8 \\ \hline \end{array}$$
$$\begin{array}{r} \$\ .61 \\ \times\ \ 9 \\ \hline \end{array}$$

2.
$$\begin{array}{r} \$\ .46 \\ \times\ \ 9 \\ \hline \end{array}$$
$$\begin{array}{r} \$\ .48 \\ \times\ \ 8 \\ \hline \end{array}$$
$$\begin{array}{r} \$\ .57 \\ \times\ \ 8 \\ \hline \end{array}$$
$$\begin{array}{r} \$\ .24 \\ \times\ \ 9 \\ \hline \end{array}$$

3.
$$\begin{array}{r} \$\ .29 \\ \times\ \ 9 \\ \hline \end{array}$$
$$\begin{array}{r} \$\ .76 \\ \times\ \ 9 \\ \hline \end{array}$$
$$\begin{array}{r} \$\ .69 \\ \times\ \ 8 \\ \hline \end{array}$$
$$\begin{array}{r} \$\ .53 \\ \times\ \ 8 \\ \hline \end{array}$$

4.
$$\begin{array}{r} \$\ .48 \\ \times\ \ 9 \\ \hline \end{array}$$
$$\begin{array}{r} \$\ .56 \\ \times\ \ 8 \\ \hline \end{array}$$
$$\begin{array}{r} \$\ .72 \\ \times\ \ 8 \\ \hline \end{array}$$
$$\begin{array}{r} \$\ .65 \\ \times\ \ 9 \\ \hline \end{array}$$

5.
$$\begin{array}{r} \$\ .85 \\ \times\ \ 8 \\ \hline \end{array}$$
$$\begin{array}{r} \$\ .41 \\ \times\ \ 9 \\ \hline \end{array}$$
$$\begin{array}{r} \$\ .76 \\ \times\ \ 8 \\ \hline \end{array}$$
$$\begin{array}{r} \$\ .95 \\ \times\ \ 8 \\ \hline \end{array}$$

6.
$$\begin{array}{r} \$\ .59 \\ \times\ \ 8 \\ \hline \end{array}$$
$$\begin{array}{r} \$\ .92 \\ \times\ \ 8 \\ \hline \end{array}$$
$$\begin{array}{r} \$\ .96 \\ \times\ \ 9 \\ \hline \end{array}$$
$$\begin{array}{r} \$\ .84 \\ \times\ \ 9 \\ \hline \end{array}$$

Name _____

Multiplying 3-digit numbers by 1 digit (2's, 3's, 4's, 5's)

Example:

$$\begin{array}{r} 2 \\ \$\ 2.05 \\ \times\quad 4 \\ \hline \$\ 8.20 \end{array}$$

1.

$$\begin{array}{r} \$\ 4.11 \\ \times\quad 3 \\ \hline \end{array}$$
$$\begin{array}{r} \$\ 2.17 \\ \times\quad 4 \\ \hline \end{array}$$
$$\begin{array}{r} \$\ 6.15 \\ \times\quad 2 \\ \hline \end{array}$$
$$\begin{array}{r} \$\ 3.87 \\ \times\quad 3 \\ \hline \end{array}$$

2.

$$\begin{array}{r} \$\ 8.14 \\ \times\quad 3 \\ \hline \end{array}$$
$$\begin{array}{r} \$\ 8.05 \\ \times\quad 5 \\ \hline \end{array}$$
$$\begin{array}{r} \$\ 2.75 \\ \times\quad 4 \\ \hline \end{array}$$
$$\begin{array}{r} \$\ 5.43 \\ \times\quad 2 \\ \hline \end{array}$$

3.

$$\begin{array}{r} \$\ 9.21 \\ \times\quad 5 \\ \hline \end{array}$$
$$\begin{array}{r} \$\ 6.03 \\ \times\quad 4 \\ \hline \end{array}$$
$$\begin{array}{r} \$\ 3.62 \\ \times\quad 4 \\ \hline \end{array}$$
$$\begin{array}{r} \$\ 6.16 \\ \times\quad 5 \\ \hline \end{array}$$

4.

$$\begin{array}{r} \$\ 6.86 \\ \times\quad 5 \\ \hline \end{array}$$
$$\begin{array}{r} \$\ 6.02 \\ \times\quad 2 \\ \hline \end{array}$$
$$\begin{array}{r} \$\ 2.12 \\ \times\quad 5 \\ \hline \end{array}$$
$$\begin{array}{r} \$\ 4.19 \\ \times\quad 3 \\ \hline \end{array}$$

5.

$$\begin{array}{r} \$\ 2.09 \\ \times\quad 4 \\ \hline \end{array}$$
$$\begin{array}{r} \$\ 2.15 \\ \times\quad 3 \\ \hline \end{array}$$
$$\begin{array}{r} \$\ 6.18 \\ \times\quad 4 \\ \hline \end{array}$$
$$\begin{array}{r} \$\ 3.04 \\ \times\quad 5 \\ \hline \end{array}$$

6.

$$\begin{array}{r} \$\ 9.03 \\ \times\quad 2 \\ \hline \end{array}$$
$$\begin{array}{r} \$\ 9.16 \\ \times\quad 4 \\ \hline \end{array}$$
$$\begin{array}{r} \$\ 6.12 \\ \times\quad 5 \\ \hline \end{array}$$
$$\begin{array}{r} \$\ 2.09 \\ \times\quad 3 \\ \hline \end{array}$$

MENU-MATH/ICE CREAM PARLOR

Multiplying 3-digit numbers by 1 digit (6's, 7's)

1.

$$\begin{array}{r} \$\ 2.15 \\ \times\quad 7 \\ \hline \end{array}\qquad \begin{array}{r} \$\ 2.98 \\ \times\quad 6 \\ \hline \end{array}\qquad \begin{array}{r} \$\ 7.46 \\ \times\quad 7 \\ \hline \end{array}\qquad \begin{array}{r} \$\ 2.67 \\ \times\quad 6 \\ \hline \end{array}$$

2.

$$\begin{array}{r} \$\ 2.42 \\ \times\quad 7 \\ \hline \end{array}\qquad \begin{array}{r} \$\ 3.43 \\ \times\quad 6 \\ \hline \end{array}\qquad \begin{array}{r} \$\ 8.28 \\ \times\quad 7 \\ \hline \end{array}\qquad \begin{array}{r} \$\ 8.12 \\ \times\quad 6 \\ \hline \end{array}$$

3.

$$\begin{array}{r} \$\ 2.56 \\ \times\quad 7 \\ \hline \end{array}\qquad \begin{array}{r} \$\ 3.58 \\ \times\quad 6 \\ \hline \end{array}\qquad \begin{array}{r} \$\ 7.56 \\ \times\quad 7 \\ \hline \end{array}\qquad \begin{array}{r} \$\ 3.23 \\ \times\quad 6 \\ \hline \end{array}$$

4.

$$\begin{array}{r} \$\ 4.69 \\ \times\quad 7 \\ \hline \end{array}\qquad \begin{array}{r} \$\ 5.03 \\ \times\quad 6 \\ \hline \end{array}\qquad \begin{array}{r} \$\ 4.63 \\ \times\quad 7 \\ \hline \end{array}\qquad \begin{array}{r} \$\ 8.56 \\ \times\quad 6 \\ \hline \end{array}$$

5.

$$\begin{array}{r} \$\ 4.75 \\ \times\quad 7 \\ \hline \end{array}\qquad \begin{array}{r} \$\ 6.98 \\ \times\quad 6 \\ \hline \end{array}\qquad \begin{array}{r} \$\ 9.15 \\ \times\quad 6 \\ \hline \end{array}\qquad \begin{array}{r} \$\ 9.28 \\ \times\quad 7 \\ \hline \end{array}$$

6.

$$\begin{array}{r} \$\ 6.10 \\ \times\quad 7 \\ \hline \end{array}\qquad \begin{array}{r} \$\ 7.12 \\ \times\quad 7 \\ \hline \end{array}\qquad \begin{array}{r} \$\ 9.07 \\ \times\quad 7 \\ \hline \end{array}\qquad \begin{array}{r} \$\ 9.93 \\ \times\quad 6 \\ \hline \end{array}$$

Name _____

1.

$ 2.48 $ 2.30 $ 6.72 $ 7.18
×   8   ×   9   ×   8   ×   9

2.

$ 1.62 $ 2.98 $ 7.23 $ 7.56
×   9   ×   8   ×   9   ×   8

3.

$ 3.42 $ 3.60 $ 8.12 $ 8.50
×   9   ×   8   ×   9   ×   8

4.

$ 4.15 $ 4.28 $ 8.75 $ 9.12
×   9   ×   8   ×   9   ×   8

5.

$ 4.73 $ 5.13 $ 9.18 $ 9.50
×   9   ×   9   ×   9   ×   8

6.

$ 5.88 $ 6.12 $ 9.63 $ 9.92
×   8   ×   9   ×   8   ×   9

Name _____

Multiplying 4-digit numbers by 1 digit (2's, 3's, 4's, 5's)

Example:
$$\begin{array}{r} \overset{1\ \ 2}{\$\ 13.15} \\ \times\quad 4 \\ \hline \$\ 52.60 \end{array}$$

1.

$$\begin{array}{r} \$\ \ 13.05 \\ \times\quad 2 \\ \hline \end{array}$$
$$\begin{array}{r} \$\ \ 14.29 \\ \times\quad 4 \\ \hline \end{array}$$
$$\begin{array}{r} \$\ \ 19.23 \\ \times\quad 2 \\ \hline \end{array}$$
$$\begin{array}{r} \$\ \ 15.10 \\ \times\quad 3 \\ \hline \end{array}$$

2.

$$\begin{array}{r} \$\ \ 20.50 \\ \times\quad 3 \\ \hline \end{array}$$
$$\begin{array}{r} \$\ \ 10.50 \\ \times\quad 4 \\ \hline \end{array}$$
$$\begin{array}{r} \$\ \ 22.52 \\ \times\quad 4 \\ \hline \end{array}$$
$$\begin{array}{r} \$\ \ 26.72 \\ \times\quad 5 \\ \hline \end{array}$$

3.

$$\begin{array}{r} \$\ \ 32.45 \\ \times\quad 4 \\ \hline \end{array}$$
$$\begin{array}{r} \$\ \ 18.39 \\ \times\quad 3 \\ \hline \end{array}$$
$$\begin{array}{r} \$\ \ 18.56 \\ \times\quad 2 \\ \hline \end{array}$$
$$\begin{array}{r} \$\ \ 16.55 \\ \times\quad 4 \\ \hline \end{array}$$

4.

$$\begin{array}{r} \$\ \ 23.52 \\ \times\quad 4 \\ \hline \end{array}$$
$$\begin{array}{r} \$\ \ 11.95 \\ \times\quad 3 \\ \hline \end{array}$$
$$\begin{array}{r} \$\ \ 12.29 \\ \times\quad 3 \\ \hline \end{array}$$
$$\begin{array}{r} \$\ \ 18.28 \\ \times\quad 5 \\ \hline \end{array}$$

5.

$$\begin{array}{r} \$\ \ 16.13 \\ \times\quad 4 \\ \hline \end{array}$$
$$\begin{array}{r} \$\ \ 35.16 \\ \times\quad 2 \\ \hline \end{array}$$
$$\begin{array}{r} \$\ \ 22.86 \\ \times\quad 3 \\ \hline \end{array}$$
$$\begin{array}{r} \$\ \ 25.66 \\ \times\quad 5 \\ \hline \end{array}$$

6.

$$\begin{array}{r} \$\ \ 24.83 \\ \times\quad 2 \\ \hline \end{array}$$
$$\begin{array}{r} \$\ \ 25.61 \\ \times\quad 4 \\ \hline \end{array}$$
$$\begin{array}{r} \$\ \ 18.49 \\ \times\quad 3 \\ \hline \end{array}$$
$$\begin{array}{r} \$\ \ 26.97 \\ \times\quad 3 \\ \hline \end{array}$$

MENU-MATH/ICE CREAM PARLOR

Name _____

1.

$\begin{array}{r} \$\ 12.52 \\ \times \quad 6 \\ \hline \end{array}$
$\begin{array}{r} \$\ 12.53 \\ \times \quad 7 \\ \hline \end{array}$
$\begin{array}{r} \$\ 25.51 \\ \times \quad 7 \\ \hline \end{array}$
$\begin{array}{r} \$\ 52.18 \\ \times \quad 6 \\ \hline \end{array}$

2.

$\begin{array}{r} \$\ 11.25 \\ \times \quad 6 \\ \hline \end{array}$
$\begin{array}{r} \$\ 11.95 \\ \times \quad 7 \\ \hline \end{array}$
$\begin{array}{r} \$\ 18.13 \\ \times \quad 7 \\ \hline \end{array}$
$\begin{array}{r} \$\ 19.63 \\ \times \quad 6 \\ \hline \end{array}$

3.

$\begin{array}{r} \$\ 24.85 \\ \times \quad 7 \\ \hline \end{array}$
$\begin{array}{r} \$\ 35.67 \\ \times \quad 7 \\ \hline \end{array}$
$\begin{array}{r} \$\ 32.44 \\ \times \quad 6 \\ \hline \end{array}$
$\begin{array}{r} \$\ 27.62 \\ \times \quad 6 \\ \hline \end{array}$

4.

$\begin{array}{r} \$\ 23.58 \\ \times \quad 6 \\ \hline \end{array}$
$\begin{array}{r} \$\ 37.89 \\ \times \quad 7 \\ \hline \end{array}$
$\begin{array}{r} \$\ 23.18 \\ \times \quad 6 \\ \hline \end{array}$
$\begin{array}{r} \$\ 25.63 \\ \times \quad 7 \\ \hline \end{array}$

5.

$\begin{array}{r} \$\ 45.23 \\ \times \quad 6 \\ \hline \end{array}$
$\begin{array}{r} \$\ 42.25 \\ \times \quad 7 \\ \hline \end{array}$
$\begin{array}{r} \$\ 53.64 \\ \times \quad 7 \\ \hline \end{array}$
$\begin{array}{r} \$\ 65.14 \\ \times \quad 6 \\ \hline \end{array}$

6.

$\begin{array}{r} \$\ 73.25 \\ \times \quad 6 \\ \hline \end{array}$
$\begin{array}{r} \$\ 81.25 \\ \times \quad 7 \\ \hline \end{array}$
$\begin{array}{r} \$\ 92.67 \\ \times \quad 6 \\ \hline \end{array}$
$\begin{array}{r} \$\ 62.59 \\ \times \quad 7 \\ \hline \end{array}$

Name _____

Multiplying 4-digit numbers by 1 digit (8's, 9's)

1.
$ 32.23
× 9

$ 35.62
× 8

$ 71.26
× 9

$ 69.12
× 8

2.
$ 32.43
× 8

$ 41.25
× 9

$ 73.45
× 9

$ 65.25
× 8

3.
$ 33.64
× 8

$ 34.92
× 9

$ 72.36
× 9

$ 64.37
× 8

4.
$ 36.23
× 8

$ 57.91
× 9

$ 73.68
× 8

$ 61.85
× 9

5.
$ 46.21
× 8

$ 47.45
× 9

$ 82.39
× 8

$ 93.56
× 9

6.
$ 26.34
× 8

$ 18.96
× 9

$ 86.44
× 8

$ 96.24
× 9

MENU-MATH/ICE CREAM PARLOR

Name _____

Multiplying 2-digit numbers by 2 digits. 3-digit answer

Example:

```
        1
$   .25
×   12
    50
  2 5
$ 3.00
```

1.
```
$   .22
×   23
```
```
$   .26
×   12
```
```
$   .33
×   14
```
```
$   .21
×   27
```

2.
```
$   .26
×   22
```
```
$   .21
×   43
```
```
$   .34
×   21
```
```
$   .34
×   25
```

3.
```
$   .23
×   36
```
```
$   .16
×   61
```
```
$   .24
×   12
```
```
$   .29
×   19
```

4.
```
$   .52
×   13
```
```
$   .43
×   12
```
```
$   .24
×   31
```
```
$   .53
×   13
```

5.
```
$   .23
×   13
```
```
$   .24
×   12
```
```
$   .62
×   13
```
```
$   .54
×   11
```

MENU-MATH/ICE CREAM PARLOR

Name _____

Multiplying 2-digit numbers by 2 digits.
4-digit answer

1. $ .91 $ .73 $ .89 $ .68
    × 23   × 16   × 66   × 37

2. $ .94 $ .44 $ .84 $ .56
    × 48   × 57   × 83   × 31

3. $ .91 $ .39 $ .46 $ .27
    × 66   × 82   × 28   × 56

4. $ .83 $ .49 $ .78 $ .86
    × 16   × 26   × 52   × 68

5. $ .78 $ .38 $ .52 $ .73
    × 43   × 46   × 31   × 32

MENU-MATH/ICE CREAM PARLOR

Name _____

Multiplying 2-digit numbers by 2 digits. 4-digit answer

1. $ .91 $ .57 $ .97 $ .83
        × 76   × 19   × 21   × 22

2. $ .82 $ .82 $ .65 $ .92
        × 51   × 42   × 71   × 24

3. $ .65 $ .84 $ .79 $ .89
        × 59   × 55   × 84   × 63

4. $ .68 $ .23 $ .72 $ .63
        × 47   × 87   × 45   × 82

5. $ .94 $ .82 $ .42 $ .36
        × 21   × 47   × 93   × 31

MENU-MATH/ICE CREAM PARLOR

Name _____

Multiplying 3-digit numbers by 2 digits. 4-digit answer

Example:

```
        2
$   6.27
×    13
   18 81
   62 7
$ 81.51
```

1.
```
$   4.12
×    24
```
```
$   4.23
×    12
```
```
$   2.12
×    23
```
```
$   1.21
×    49
```

2.
```
$   1.32
×    24
```
```
$   1.31
×    27
```
```
$   1.21
×    57
```
```
$   4.25
×    16
```

3.
```
$   1.23
×    32
```
```
$   5.27
×    15
```
```
$   3.24
×    26
```
```
$   1.32
×    62
```

4.
```
$   3.14
×    21
```
```
$   2.23
×    32
```
```
$   5.27
×    15
```
```
$   1.28
×    52
```

5.
```
$   2.86
×    14
```
```
$   1.53
×    18
```
```
$   4.22
×    17
```
```
$   6.27
×    13
```

MENU-MATH/ICE CREAM PARLOR

Name _____

Multiplying 3-digit numbers by 2 digits. 5-digit answer

1.	$ 8.25 × 93	$ 4.64 × 66	$ 9.26 × 93	$ 5.89 × 63

1.　　$ 8.25　　　$ 4.64　　　$ 9.26　　　$ 5.89
　　　×　93　　　×　66　　　×　93　　　×　63

2.　　$ 6.48　　　$ 9.87　　　$ 9.27　　　$ 3.21
　　　×　24　　　×　73　　　×　47　　　×　73

3.　　$ 4.23　　　$ 6.82　　　$ 3.69　　　$ 4.28
　　　×　29　　　×　57　　　×　84　　　×　96

4.　　$ 7.84　　　$ 3.57　　　$ 2.73　　　$ 5.37
　　　×　72　　　×　85　　　×　83　　　×　49

5.　　$ 5.62　　　$ 8.69　　　$ 3.27　　　$ 4.86
　　　×　62　　　×　36　　　×　48　　　×　94

MENU-MATH/ICE CREAM PARLOR

Name _____

Multiplying 3-digit numbers by 2 digits. 5-digit answer

1. $ 7.29 $ 4.28 $ 9.24 $ 4.78
    × 84   × 76   × 27   × 82

2. $ 3.75 $ 3.42 $ 2.39 $ 3.71
    × 89   × 82   × 51   × 54

3. $ 5.79 $ 8.18 $ 5.48 $ 5.78
    × 32   × 49   × 91   × 43

4. $ 8.49 $ 7.21 $ 2.96 $ 9.37
    × 82   × 52   × 35   × 83

5. $ 5.63 $ 9.39 $ 9.19 $ 9.39
    × 28   × 75   × 59   × 49

Name _____

**Multiplying 3-digit
numbers by 2 digits,
zero difficulty.
5-digit answer**

---

1. $ 9.02 $ 7.04 $ 6.08 $ 4.04
    ×   38   ×   27   ×   34   ×   84

2. $ 6.01 $ 4.05 $ 6.07 $ 7.08
    ×   98   ×   86   ×   45   ×   72

3. $ 3.05 $ 4.05 $ 7.07 $ 7.05
    ×   53   ×   72   ×   20   ×   80

4. $ 8.07 $ 6.09 $ 5.04 $ 7.06
    ×   69   ×   38   ×   41   ×   65

5. $ 6.30 $ 7.06 $ 6.09 $ 8.07
    ×   82   ×   60   ×   50   ×   17

19   MENU-MATH/ICE CREAM PARLOR

Name _____

Example:
```
      6
    1 1 2
$  39.24
×     72
---------
   78 48
  2746 8
---------
$ 2825.28
```

1.
```
$  83.42
×     25
```
```
$  63.18
×     41
```
```
$  58.66
×     32
```
```
$  89.25
×     43
```

2.
```
$  52.44
×     38
```
```
$  86.85
×     33
```
```
$  29.18
×     79
```
```
$  66.58
×     27
```

3.
```
$  41.29
×     82
```
```
$  66.13
×     49
```
```
$  82.36
×     84
```
```
$  29.46
×     52
```

4.
```
$  91.88
×     26
```
```
$  44.18
×     32
```
```
$  85.88
×     62
```
```
$  71.82
×     93
```

5.
```
$  46.62
×     31
```
```
$  68.85
×     82
```
```
$  34.43
×     64
```
```
$  86.86
×     64
```

Name _____

Multiplying 4-digit numbers by 2 digits.
6-digit answer

1. $ 59.63 $ 87.57 $ 58.96 $ 56.85
    ×   57   ×   35   ×   24   ×   76

2. $ 87.56 $ 67.86 $ 77.85 $ 49.73
    ×   38   ×   29   ×   63   ×   35

3. $ 87.31 $ 53.28 $ 78.95 $ 69.62
    ×   72   ×   47   ×   49   ×   81

4. $ 75.00 $ 89.00 $ 76.00 $ 58.00
    ×   53   ×   78   ×   74   ×   69

5. $ 62.00 $ 19.00 $ 36.00 $ 56.70
    ×   58   ×   98   ×   52   ×   96

MENU-MATH/ICE CREAM PARLOR

Name _____

**Multiplying 4-digit
numbers by 2 digits,
zero difficulty.
5- and 6-digit answers**

1. $ 20.05 $ 32.08 $ 80.04 $ 20.03
    ×   62   ×   16   ×   91   ×   62

2. $ 70.03 $ 60.04 $ 20.04 $ 31.08
    ×   42   ×   15   ×   27   ×   52

3. $ 60.01 $ 21.08 $ 50.05 $ 42.07
    ×   82   ×   64   ×   25   ×   36

4. $ 30.60 $ 40.20 $ 20.90 $ 80.20
    ×   59   ×   88   ×   66   ×   29

5. $ 60.30 $ 50.08 $ 20.40 $ 80.30
    ×   12   ×   17   ×   56   ×   59

MENU-MATH/ICE CREAM PARLOR

Finding Sales Tax

Some states charge sales tax. When you buy food in a restaurant, you must pay sales tax. The amount of money you pay for sales tax depends on how much the food costs. Look carefully at the example below.

Ice Cream Parlor

2	White Cakes	$ 5.40
1	Fancy Sundae	2.60
	Subtotal	$ 8.00
	Tax	.56
	Total	$ 8.56

Directions:
Follow the steps below.

1. Add the two items to find the **subtotal**.

2. Write in the **tax** from the Tax Table.

3. Add the **subtotal** and the **tax** to find the **total**.

You will be practicing with a **7% Sales Tax Table.** Find out from your teacher what tax rate your state charges.

The **subtotal** on the sample check above is **$8.00**
$8.00 is between $7.93 and $8.07 on the Tax Table. Tax owed is $.56.

7% Tax Table

Amount of Sale	Tax	Amount of Sale	Tax
5.50 - 5.64	.39	7.22 - 7.35	.51
5.65 - 5.78	.40	7.36 - 7.49	.52
5.79 - 5.92	.41	7.50 - 7.64	.53
5.93 - 6.07	.42	7.65 - 7.78	.54
6.08 - 6.21	.43	7.79 - 7.92	.55
6.22 - 6.35	.44	**7.93 - 8.07**	**.56**
6.36 - 6.49	.45	8.08 - 8.21	.57
6.50 - 6.64	.46	8.22 - 8.35	.58
6.65 - 6.78	.47	8.36 - 8.49	.59
6.79 - 6.92	.48	8.50 - 8.64	.60
6.93 - 7.07	.49	8.65 - 8.78	.61
7.08 - 7.21	.50	8.79 - 8.92	.62

MENU-MATH/ICE CREAM PARLOR

7% SALES TAX TABLE

Amount of Sale	Tax	Amount of Sale	Tax	Amount of Sale	Tax
14.08 - 14.21	.99	20.08 - 20.21	1.41	26.08 - 26.21	1.83
14.22 - 14.35	1.00	20.22 - 20.35	1.42	26.22 - 26.35	1.84
14.36 - 14.49	1.01	20.36 - 20.49	1.43	26.36 - 26.49	1.85
14.50 - 14.64	1.02	20.50 - 20.64	1.44	26.50 - 26.64	1.86
14.65 - 14.78	1.03	20.65 - 20.78	1.45	26.65 - 26.78	1.87
14.79 - 14.92	1.04	20.79 - 20.92	1.46	26.79 - 26.92	1.88
14.93 - 15.07	1.05	20.93 - 21.07	1.47	26.93 - 27.07	1.89
15.08 - 15.21	1.06	21.08 - 21.21	1.48	27.08 - 27.21	1.90
15.22 - 15.35	1.07	21.22 - 21.35	1.49	27.22 - 27.35	1.91
15.36 - 15.49	1.08	21.36 - 21.49	1.50	27.36 - 27.49	1.92
15.50 - 15.64	1.09	21.50 - 21.64	1.51	27.50 - 27.64	1.93
15.65 - 15.78	1.10	21.65 - 21.78	1.52	27.65 - 27.78	1.94
15.79 - 15.92	1.11	21.79 - 21.92	1.53	27.79 - 27.92	1.95
15.93 - 16.07	1.12	21.93 - 22.07	1.54	27.93 - 28.07	1.96
16.08 - 16.21	1.13	22.08 - 22.21	1.55	28.08 - 28.21	1.97
16.22 - 16.35	1.14	22.22 - 22.35	1.56	28.22 - 28.35	1.98
16.36 - 16.49	1.15	22.36 - 22.49	1.57	28.36 - 28.49	1.99
16.50 - 16.64	1.16	22.50 - 22.64	1.58	28.50 - 28.64	2.00
16.65 - 16.78	1.17	22.65 - 22.78	1.59	28.65 - 28.78	2.01
16.79 - 16.92	1.18	22.79 - 22.92	1.60	28.79 - 28.92	2.02
16.93 - 17.07	1.19	22.93 - 23.07	1.61	28.93 - 29.07	2.03
17.08 - 17.21	1.20	23.08 - 23.21	1.62	29.08 - 29.21	2.04
17.22 - 17.35	1.21	23.22 - 23.35	1.63	29.22 - 29.35	2.05
17.36 - 17.49	1.22	23.36 - 23.49	1.64	29.36 - 29.49	2.06
17.50 - 17.64	1.23	23.50 - 23.64	1.65	29.50 - 29.64	2.07
17.65 - 17.78	1.24	23.65 - 23.78	1.66	29.65 - 29.78	2.08
17.79 - 17.92	1.25	23.79 - 23.92	1.67	29.79 - 29.92	2.09
17.93 - 18.07	1.26	23.93 - 24.07	1.68	29.93 - 30.07	2.10
18.08 - 18.21	1.27	24.08 - 24.21	1.69	30.08 - 30.21	2.11
18.22 - 18.35	1.28	24.22 - 24.35	1.70	30.22 - 30.35	2.12
18.36 - 18.49	1.29	24.36 - 24.49	1.71	30.36 - 30.49	2.13
18.50 - 18.64	1.30	24.50 - 24.64	1.72	30.50 - 30.64	2.14
18.65 - 18.78	1.31	24.65 - 24.78	1.73	30.65 - 30.78	2.15
18.79 - 18.92	1.32	24.79 - 24.92	1.74	30.79 - 30.92	2.16
18.93 - 19.07	1.33	24.93 - 25.07	1.75	30.93 - 31.07	2.17
19.08 - 19.21	1.34	25.08 - 25.21	1.76	31.08 - 31.21	2.18
19.22 - 19.35	1.35	25.22 - 25.35	1.77	31.22 - 31.35	2.19
19.36 - 19.49	1.36	25.36 - 25.49	1.78	31.36 - 31.49	2.20
19.50 - 19.64	1.37	25.50 - 25.64	1.79	31.50 - 31.64	2.21
19.65 - 19.78	1.38	25.65 - 25.78	1.80	31.65 - 31.78	2.22
19.79 - 19.92	1.39	25.79 - 25.92	1.81	31.79 - 31.92	2.23
19.93 - 20.07	1.40	25.93 - 26.07	1.82	31.93 - 32.07	2.24

Name _____

Directions:
Use the tax table on page 24
to find the sales tax on the
following amounts.

	AMOUNT	SALES TAX	TOTAL
1.	$14.58	_____	_____
2.	$15.93	_____	_____
3.	$16.75	_____	_____
4.	$18.49	_____	_____
5.	$19.92	_____	_____
6.	$22.18	_____	_____
7.	$23.97	_____	_____
8.	$24.25	_____	_____
9.	$24.65	_____	_____
10.	$25.03	_____	_____
11.	$26.01	_____	_____
12.	$27.90	_____	_____
13.	$28.33	_____	_____
14.	$29.67	_____	_____
15.	$30.35	_____	_____
16.	$31.02	_____	_____

Name _____

**Proper placement of the
decimal point and
dollar sign**

When adding, subtracting, multiplying, and dividing money, always use a dollar sign
($) and a decimal point (.) in your answer.

When adding and subtracting money, line up the decimal points.

$$\begin{array}{r} \$\ 18.36 \\ +\ \ 5.23 \\ \hline \$\ 23.59 \end{array}$$ ◄──── **EXAMPLES** ────► $$\begin{array}{r} \$\ 23.62 \\ -\ \ 4.03 \\ \hline \$\ 19.59 \end{array}$$

Line up the decimal points.

When multiplying money, put the same number of decimal places in the answer as
there are in the problem.

Example:
$$\begin{array}{r} \$\ 36.25 \\ \times\ \ \ \ 26 \\ \hline 217\ 50 \\ 725\ 0\ \ \\ \hline \$\ 942.50 \end{array}$$
**2 decimal places
in the problem**

**There must be 2
decimal places
in the answer.**

When dividing money by a whole number, begin by placing the decimal point in the
answer.

Example:
$$2\ \overline{)\ \$\ 41.62}\ ^{\textstyle .}$$
**Line up the
decimal point.**

1. **ADD**	2. **SUBTRACT**
$4.06 + $3.62 + $1.25 + $.62 =	$20.05 − $1.10 =
3. **MULTIPLY**	4. **DIVIDE**
$12.15 × 6 =	$32.12 ÷ 2 =

Name _____

Multiplying to find the cost when buying more than one of the same item.

Example:
Two Royal
Banana Splits
@ $3.00 ea.

$ 3.00
×   2
$ 6.00

---

1. Five Triple Sundaes   @ $ 3.20 ea.   _____

2. Twelve regular Colas   @ $ .95 ea.   _____

3. Seven Grand Banana Splits   @ $ 4.95 ea.   _____

4. Nine Giant Sundaes   @ $ 6.95 ea.   _____

5. Twenty-two Super Sodas   @ $ 2.60 ea.   _____

6. Eighteen Apple Banana Splits   @ $ 2.80 ea.   _____

7. Twelve Rainbow Sundaes   @ $12.95 ea.   _____

8. Eight Hot Chocolates   @ $ .75 ea.   _____

9. Twenty-five Fancy Sundaes   @ $ 2.60 ea.   _____

10. Twenty-nine Special Sundaes   @ $ 2.45 ea.   _____

11. Sixteen Root Beer Floats   @ $ 2.10 ea.   _____

12. Twelve Nutty Sundaes   @ $ 3.65 ea.   _____

13. Eleven Raspberry Freezes   @ $ 2.25 ea.   _____

14. Ten Golden Cakes   @ $ 2.70 ea.   _____

MENU-MATH/ICE CREAM PARLOR

Name _____

Multiplying to find the cost when buying more than one of the same item.

Directions:

    1. **Use the menu to find the item price.**
    2. **Multiply to find the total price.**

	Item Price	Total Price
1. Six Berry Banana Splits	_____	_____
2. Ten Big Scoop Sundaes	_____	_____
3. Fifteen Hot Fudge Sundaes	_____	_____
4. Twenty-two Scotch Sundaes	_____	_____
5. Twenty-five Fudge Cakes	_____	_____
6. Thirty Cola Floats	_____	_____
7. Six Rainbow Sundaes	_____	_____
8. Eight Tin Roof Sundaes	_____	_____
9. Forty-two Raspberry Freezes	_____	_____
10. Fifty Nutty Sundaes	_____	_____
11. Thirty-three Vanilla Sodas	_____	_____
12. Sixteen Pecan Banana Splits	_____	_____
13. Eighteen American Banana Splits	_____	_____
14. Thirty-seven Fancy Sundaes	_____	_____

Name _____

Solving addition and multiplication story problems.

Directions: Use the menu to solve the following story problems (all answers exclude sales tax).

_____

1. What would be the cost of 5 Giant Sundaes?   _____

2. How much would 7 glasses of Soda Water cost?   _____

3. What would you pay for 8 Scotch Sundaes?   _____

4. Nine girls each have a small Root Beer.
   What does this cost?   _____

5. Seven people each order a Rainbow Sundae.
   How much will their check total?   _____

6. Four girls went to the Old Fashioned Ice
   Cream Parlor, and each ordered an Apple Banana
   Split. Altogether, how much did they spend?   _____

7. After the basketball game, 9 of the players
   went for ice cream. They each had a
   Jumbo Scoop of Butter-Pecan ice cream.
   What did their check total?   _____

8. Jack and his family went to the Old Fashioned
   Ice Cream Parlor. They ordered 9 lime Freezes
   and left a $3.00 tip for the waitress.
   How much did they spend?   _____

9. Chris and his friends stopped at the
   Ice Cream Parlor after the dance. They ate 4
   Berry Banana Splits and 5 Nutty Sundaes.
   How much did they spend?   _____

Name _____

Solving addition and multiplication story problems.

Directions: Use the menu to solve the following story problems (all answers exclude sales tax).

1. You and your friends order 3 Giant Sundaes and 4 Simple Sundaes. How much money will you give the cashier?

   _____

2. Mr. Strong took his children and their friends for ice cream sodas. They ordered 4 Chocolate Sodas and 4 Strawberry Sodas. How much did Mr. Strong spend?

   _____

3. Terry earned money babysitting. She took her friends for ice cream. They ordered 7 Fancy Sundaes and 3 Special Sundaes. How much did Terry spend on ice cream?

   _____

4. Your family orders 3 Tin Roof Sundaes, 6 small Colas, and 3 Super Shakes. How much does this cost?

   _____

5. Four girls each have a piece of Fudge Cake. They also each have a glass of milk. Altogether, how much money will they owe?

   _____

6. You are having a surprise birthday party for your friend at the Old Fashioned Ice Cream Parlor. You decide to invite 14 people to attend the party. The waitress suggests that you order 2 Rainbow Sundaes and 15 regular Sprites. What is the cost of the party?

   _____

7. Patty and Bob went to the Ice Cream Parlor to buy 6 pieces of White Cake, 4 Giant Sundaes, and 3 Royal Banana Splits for their class. How much will they pay the cashier?

   _____

Name _____

Solving addition, subtraction, and multiplication story problems.

Directions: Use the menu to solve the following story problems (all answers exclude sales tax).

1. The James family orders 6 American Banana Splits. If they use a $20 bill to pay, how much change will they receive? _____

2. Mrs. Casey treated each student in her class to a single scoop of ice cream. There are 15 students in her class. How much did this cost Mrs. Casey? _____

   How much change did she receive from $20.00? _____

3. Members of the Girls' Club go for ice cream. Four girls each have a Special Sundae and 5 girls each order Triple Sundaes. Altogether, how much do the sundaes cost? _____

   What change do they receive from $50.00? _____

4. Lisa, Tom, Mike, and Jill each ordered a piece of Golden Cake. Also, they each had a small Cola. How much did their check total? _____

   How much change did they receive from $20.00? _____

5. Five people order the following: 2 Berry Banana Splits, 2 Triple Sundaes, 4 regular Root Beers. What will be the total of their check? _____

   How much change will they get back from $20.00? _____

6. Betty had $46.50 to spend. She invited several of her friends to the Parlor to celebrate her birthday. They ordered 3 Grand Banana Splits, 2 Giant Sundaes, and 12 regular Sprites. How much did this cost? _____

   How much money will Betty have left? _____

Name _____

Solving multiplication story problems.

Directions: Solve the following story problems without the menu.

1. A gallon of ice cream at the Old Fashioned Ice Cream Parlor costs $5.20. What would be the cost of 15 gallons of ice cream?

   _____

2. Sara makes $4.25 an hour working at the Parlor. If she works 32 hours, how much money will she earn?

   _____

3. A giant Sundae costs $6.95 at the Old Fashioned Ice Cream Parlor. If 110 of these sundaes are sold, how much money will be made?

   _____

4. How much money will the Parlor make by selling 50 Super Shakes at $2.65 per shake?

   _____

5. Ice Cream Floats are $2.10. How much money will the Parlor collect if 75 floats are sold?

   _____

6. The Old Fashioned Ice Cream Parlor uses 15 gallons of milk a day. How many gallons will be used in 31 days?

   _____

7. The Parlor uses 25 jars of fudge sauce a week. How many jars will be used in 16 weeks?

   _____

8. A single scoop ice cream cone costs $.80. The Ice Cream Parlor sold 232 single scoop ice cream cones in one day. How much money was made?

   _____

9. The Parlor uses 32 pounds of bananas a week. How many pounds will be used in 12 weeks?

   _____

Name _____

Multiplying to find amount of money earned.

Directions:
Figure out how much each employee has earned working at the Parlor.

Example:

$$
\begin{array}{r}
\$\ 4.25 \\
\times\ \ 38 \\
\hline
34\ 00 \\
127\ 5 \\
\hline
\$161.50
\end{array}
$$

hourly wage
hours worked

amount earned

EMPLOYEES	HOURS WORKED	HOURLY WAGE	AMOUNT EARNED
1. Sherri	23	$4.30	_____
2. Chuck	25	$4.40	_____
3. Fred	18	$4.25	_____
4. Steve	34	$4.30	_____
5. Connie	27	$4.50	_____
6. Sandra	39	$4.25	_____
7. Tom	33	$4.55	_____
8. Dean	31	$4.45	_____
9. Andy	36	$4.35	_____
10. Jackie	35	$4.60	_____
11. Tim	17	$4.25	_____
12. Don	26	$4.30	_____

Name _____

Multiplying to find the cost of each item and adding to find the total (exclude sales tax).

After the football game, the team stopped at the Ice Cream Parlor. Each group placed its order separately.

Example: **1. Multiply to find cost. 2. Add to find total.**

3 Berry Banana Splits $2.80 ea. $ 2.80 $ .65 $ 8.40
3 Small Colas $ .65 ea.   ×   3   ×   3   + 1.95
                                     $ 8.40 $ 1.95 $ 10.35 **Total**

Use the menu and find the cost of the following orders.

1. 2 Triple Sundaes   _____   5. 4 Nutty Sundaes   _____
    4 Root Beer Floats   _____   4 Vanilla Sodas   _____
    6 regular Colas   _____   8 small Root Beers   _____
              Total: _____   Total: _____

2. 5 Fancy Sundaes   _____   6. 6 Chocolate Shakes   _____
    4 Root Beer Floats   _____   2 Pineapple Sodas   _____
    9 Milks   _____   8 small Sprites   _____
              Total: _____   Total: _____

3. 2 Nutty Sundaes   _____   7. 7 Raspberry Freezes   _____
    3 Fudge Cakes   _____   2 Giant Sundaes   _____
    5 Hot Chocolates   _____   9 small Colas   _____
              Total: _____   Total: _____

4. 4 Apple Banana Splits _____   8. 3 Simple Sundaes   _____
    3 Golden Cakes   _____   2 Cola Floats   _____
    7 Iced Teas   _____   5 small Sprites   _____
              Total: _____   Total: _____

                                       MENU-MATH/ICE CREAM PARLOR

Name _____

1. Use the menu to find prices.
2. Multiply to find item costs.
3. Add check to find subtotal.

4. Use tax table to find amount of tax.
5. Add the subtotal and tax.
6. Subtract total from amount paid.

Ice Cream Parlor		
4	Scotch Sundaes	
3	small Root Beers	
	Subtotal	
	Tax	
	Total	

Amount Paid: $20.00   Change: _____

Ice Cream Parlor		
3	Grand Banana Splits	
3	Milks	
	Subtotal	
	Tax	
	Total	

Amount Paid: $20.00   Change: _____

Ice Cream Parlor		
5	Orange Freezes	
4	Cola Floats	
	Subtotal	
	Tax	
	Total	

Amount Paid: $25.00   Change: _____

Ice Cream Parlor		
4	White Cakes	
3	Big Scoop Sundaes	
	Subtotal	
	Tax	
	Total	

Amount Paid: $30.00   Change: _____

MENU-MATH/ICE CREAM PARLOR

Name _____

1. Use the menu to find prices.
2. Multiply to find item costs.
3. Add check to find subtotal.

4. Use tax table to find amount of tax.
5. Add the subtotal and tax.
6. Subtract total from amount paid.

Ice Cream Parlor

2	Big Scoop Sundaes	
1	American Banana Split	
3	regular Colas	
3	Golden Cakes	
	Subtotal	
	Tax	
	Total	

Amount Paid: $25.00   Change: _____

Ice Cream Parlor

3	Simple Sundaes	
2	Strawberry Shakes	
1	Lime Freeze	
	Subtotal	
	Tax	
	Total	

Amount Paid: $21.00   Change: _____

Ice Cream Parlor

2	Giant Sundaes	
2	White Cakes	
3	small Sprites	
	Subtotal	
	Tax	
	Total	

Amount Paid: $30.00   Change: _____

Ice Cream Parlor

2	Apple Banana Splits	
3	Special Sundaes	
2	Nutty Sundaes	
	Subtotal	
	Tax	
	Total	

Amount Paid: $25.00   Change: _____

Name _____

1. Use the menu to find prices.
2. Multiply to find item costs.
3. Add check to find subtotal.
4. Use tax table to find amount of tax.
5. Add the subtotal and tax.
6. Subtract total from amount paid.

Ice Cream Parlor		
2	Giant Sundaes	
3	Nutty Sundaes	
2	Root Beer Floats	
Subtotal		
Tax		
Total		

Amount Paid: $40.00   Change: _____

Ice Cream Parlor		
3	Grand Banana Splits	
3	Berry Banana Splits	
2	small Colas	
3	small Root Beers	
Subtotal		
Tax		
Total		

Amount Paid: $50.00   Change: _____

Ice Cream Parlor		
3	Vanilla Super Sodas	
3	Cherry Shakes	
3	Fancy Sundaes	
Subtotal		
Tax		
Total		

Amount Paid: $30.00   Change: _____

Ice Cream Parlor		
3	Fudge Cakes	
2	Golden Cakes	
5	Coffees	
Subtotal		
Tax		
Total		

Amount Paid: $20.00   Change: _____

Name _____

1. Use the menu to find prices.
2. Multiply to find item costs.
3. Add check to find subtotal.

4. Use tax table to find amount of tax.
5. Add the subtotal and tax.
6. Subtract total from amount paid.

Ice Cream Parlor		
3	Giant Sundaes	
9	regular Colas	
	Subtotal	
	Tax	
	Total	

Amount Paid: $35.00   Change: _____

Ice Cream Parlor		
4	Royal Banana Splits	
5	Coffees	
	Subtotal	
	Tax	
	Total	

Amount Paid: $20.00   Change: _____

Ice Cream Parlor		
4	Triple Sundaes	
2	Special Sundaes	
6	small Colas	
	Subtotal	
	Tax	
	Total	

Amount Paid: $25.00   Change: _____

Ice Cream Parlor		
3	Lime Freezes	
4	Fancy Sundaes	
8	Milks	
	Subtotal	
	Tax	
	Total	

Amount Paid: $30.00   Change: _____

Name _____

1. Use the menu to find prices.
2. Multiply to find item costs.
3. Add check to find subtotal.

4. Use tax table to find amount of tax.
5. Add the subtotal and tax.
6. Subtract total from amount paid.

Ice Cream Parlor

4	Fudge Cakes	
4	Chocolate Super Sodas	
3	Strawberry Shakes	
5	small Root Beers	
	Subtotal	
	Tax	
	Total	

Amount Paid: $40.00   Change: _____

Ice Cream Parlor

5	Hot Teas	
6	small Colas	
3	Apple Banana Splits	
2	White Cakes	
	Subtotal	
	Tax	
	Total	

Amount Paid: $30.00   Change: _____

Ice Cream Parlor

3	Grand Banana Splits	
5	Vanilla Shakes	
2	Raspberry Freezes	
	Subtotal	
	Tax	
	Total	

Amount Paid: $50.00   Change: _____

Ice Cream Parlor

5	Tin Roof Sundaes	
5	small Sprites	
4	Coffees	
	Subtotal	
	Tax	
	Total	

Amount Paid: $25.00   Change: _____

Name _____

1. Use the menu to find prices.
2. Multiply to find item costs.
3. Add check to find subtotal.

4. Use tax table to find amount of tax.
5. Add the subtotal and tax.
6. Subtract total from amount paid.

	Ice Cream Parlor	
3	Fancy Sundaes	
3	Super Shakes with Malt	
4	Iced Teas	
1	Rainbow Sundae	
	Subtotal	
	Tax	
	Total	

Amount Paid: $40.00   Change: _____

	Ice Cream Parlor	
6	Golden Cakes	
3	Fudge Cakes	
8	Coffees	
	Subtotal	
	Tax	
	Total	

Amount Paid: $35.00   Change: _____

	Ice Cream Parlor	
2	Grand Banana Splits	
4	Tin Roof Sundaes	
2	small Colas	
2	Root Beer Floats	
1	Royal Banana Split	
	Subtotal	
	Tax	
	Total	

Amount Paid: $40.00   Change: _____

	Ice Cream Parlor	
7	Cola Super Floats	
6	Single Scoops	
7	small Sprites	
5	regular Colas	
	Subtotal	
	Tax	
	Total	

Amount Paid: $50.00   Change: _____

Name _____

POST TEST — MULTIPLICATON
(answers exclude sales tax)

1.
$$\$ \ .76 \times 5 \qquad \$ \ .54 \times 6 \qquad \$ \ .38 \times 9 \qquad \$ \ 6.19 \times 4 \qquad \$ \ 9.19 \times 7$$

2.
$$\$ \ 6.25 \times 8 \qquad \$ \ 13.53 \times 4 \qquad \$ \ 62.31 \times 7 \qquad \$ \ 48.52 \times 9$$

3.
$$\$ \ .93 \times 28 \qquad \$ \ .79 \times 82 \qquad \$ \ 5.29 \times 13 \qquad \$ \ 70.24 \times 16 \qquad \$ \ 60.00 \times 25$$

4. Six Triple Sundaes @ $ 3.20 ea. _____

5. Twenty-three Apple Banana Splits @ $ 2.80 ea. _____

6. Thirty Rainbow Sundaes @ $12.95 ea. _____

7. You and two friends go to the Old Fashioned Ice Cream
Parlor. You each order Grand Banana Splits which are
$4.95 apiece. How much is your bill? _____

8. Rick needs $950.00 more to buy the car he wants.
He found a job at the Ice Cream Parlor. He makes $4.80
an hour. Last month he worked 98 hours.
How much did he make? _____

How much more does he need to buy the car? _____

Name _____

Dividing 2-digit numbers by 1 digit, no remainder.

$$\begin{array}{r} \$\ .18 \\ 5\ \overline{\big)\ \$\ .90} \\ \underline{5} \\ 40 \\ \underline{40} \end{array}$$

1.
   2 ⟌ $ .42   3 ⟌ $ .36   4 ⟌ $ .64   3 ⟌ $ .72

2.
   6 ⟌ $ .96   5 ⟌ $ .75   3 ⟌ $ .87   8 ⟌ $ .96

3.
   5 ⟌ $ .90   4 ⟌ $ .56   2 ⟌ $ .96   4 ⟌ $ .68

4.
   3 ⟌ $ .48   9 ⟌ $ .90   6 ⟌ $ .66   4 ⟌ $ .84

5.
   7 ⟌ $ .98   5 ⟌ $ .65   3 ⟌ $ .75   3 ⟌ $ .96

Name _____

**Dividing 2-digit
numbers by 1 digit,
no remainder.**

1.
4 | $ .72   4 | $ .48   2 | $ .84   3 | $ .51

2.
5 | $ .95   3 | $ .63   6 | $ .90   2 | $ .60

3.
3 | $ .96   2 | $ .64   8 | $ .80   9 | $ .99

4.
3 | $ .39   2 | $ .64   7 | $ .77   6 | $ .78

5.
4 | $ .60   6 | $ .72   5 | $ .80   8 | $ .88

43   **MENU-MATH/ICE CREAM PARLOR**

Name _____

**Dividing 2-digit
numbers by 1 digit,
remainder.**

EXAMPLE:

$$\begin{array}{r} \$\ .18 \ \text{R} \ 1¢ \\ 4\ \overline{)\ \$\ .73} \\ \underline{4} \\ 33 \\ \underline{32} \\ 1 \end{array}$$

1.
3 | $ .92   2 | $ .49   3 | $ .37   6 | $ .70

2.
8 | $ .95   5 | $ .52   7 | $ .86   2 | $ .87

3.
8 | $ .91   7 | $ .96   9 | $ .98   5 | $ .82

4.
2 | $ .57   3 | $ .80   4 | $ .89   7 | $ .79

5.
4 | $ .81   2 | $ .71   3 | $ .88   6 | $ .81

Name _____

Dividing 3-digit numbers by 1 digit, no remainder.

EXAMPLE:

```
        $ 2.08
  4  |  $ 8.32
        8
      ------
        3
        0
      ------
        32
        32
```

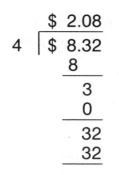

1.

2 | $ 8.10   8 | $ 8.00   2 | $ 6.48   5 | $ 8.05

2.

4 | $ 8.32   6 | $ 9.60   3 | $ 8.13   9 | $ 9.99

3.

2 | $ 3.46   4 | $ 4.76   6 | $ 7.26   3 | $ 4.29

4.

7 | $ 8.47   8 | $ 9.60   6 | $ 7.86   5 | $ 7.55

Name _____

Dividing 3-digit numbers by 1 digit, remainder.

EXAMPLE:

$$\begin{array}{r} \$\ 4.23 \text{ R } 1\text{¢} \\ 2\ \overline{)\ \$\ 8.47} \\ \underline{8} \\ 4 \\ \underline{4} \\ 7 \\ \underline{6} \\ 1 \end{array}$$

1.

8 $\overline{)\ \$\ 3.80}$ 4 $\overline{)\ \$\ 8.43}$ 5 $\overline{)\ \$\ 3.29}$ 7 $\overline{)\ \$\ 7.94}$

2.

8 $\overline{)\ \$\ 5.24}$ 3 $\overline{)\ \$\ 6.98}$ 9 $\overline{)\ \$\ 1.85}$ 6 $\overline{)\ \$\ 8.49}$

3.

4 $\overline{)\ \$\ 3.93}$ 2 $\overline{)\ \$\ 5.77}$ 3 $\overline{)\ \$\ 6.23}$ 7 $\overline{)\ \$\ 9.09}$

4.

5 $\overline{)\ \$\ 4.69}$ 6 $\overline{)\ \$\ 5.37}$ 2 $\overline{)\ \$\ 7.27}$ 8 $\overline{)\ \$\ 4.55}$

Name _____

**Dividing 3-digit
numbers by 1 digit,
with and without
remainders.**

1.

3 | $ 1.95   5 | $ 2.53   2 | $ 1.78   4 | $ 3.42

2.

7 | $ 1.54   9 | $ 7.22   3 | $ 2.25   2 | $ 6.01

3.

7 | $ 2.10   5 | $ 5.28   5 | $ 8.00   9 | $ 9.95

4.

8 | $ 8.96   3 | $ 7.29   4 | $ 7.05   2 | $ 8.53

Name _____

Dividing 4-digit numbers by 1 digit, no remainder.

EXAMPLE:

$$
\begin{array}{r}
\$32.00 \\
3\,\overline{)\$96.00} \\
9 \\
\hline
6 \\
6 \\
\hline
00
\end{array}
$$

1.
$$3\,\overline{)\$99.00}\qquad 6\,\overline{)\$66.00}\qquad 2\,\overline{)\$44.00}\qquad 3\,\overline{)\$66.00}$$

2.
$$9\,\overline{)\$90.00}\qquad 2\,\overline{)\$88.00}\qquad 3\,\overline{)\$63.00}\qquad 3\,\overline{)\$36.00}$$

3.
$$8\,\overline{)\$24.00}\qquad 5\,\overline{)\$55.00}\qquad 6\,\overline{)\$36.00}\qquad 5\,\overline{)\$40.00}$$

4.
$$7\,\overline{)\$35.00}\qquad 6\,\overline{)\$42.00}\qquad 2\,\overline{)\$46.00}\qquad 9\,\overline{)\$72.00}$$

Name _____

Dividing 4-digit numbers by 1 digit, no remainder.

1.
$$4\overline{)\,\$70.84\,}$$
$$3\overline{)\,\$59.16\,}$$
$$6\overline{)\,\$82.86\,}$$
$$2\overline{)\,\$75.68\,}$$

2.
$$7\overline{)\,\$95.97\,}$$
$$5\overline{)\,\$98.75\,}$$
$$7\overline{)\,\$42.91\,}$$
$$3\overline{)\,\$25.80\,}$$

3.
$$6\overline{)\,\$33.48\,}$$
$$5\overline{)\,\$37.45\,}$$
$$3\overline{)\,\$29.22\,}$$
$$5\overline{)\,\$43.50\,}$$

4.
$$9\overline{)\,\$62.55\,}$$
$$2\overline{)\,\$12.48\,}$$
$$7\overline{)\,\$35.63\,}$$
$$4\overline{)\,\$82.56\,}$$

© 1980, 1991 REMEDIA PUBLICATIONS

MENU-MATH/ICE CREAM PARLOR

Name _____

Dividing 4-digit numbers by 1 digit, remainder.

EXAMPLE:

$$
\begin{array}{r}
\$\ 7.68 \text{ R } 1¢ \\
3\ \overline{)\ \$23.05} \\
21 \\
\hline
2\ 0 \\
1\ 8 \\
\hline
25 \\
24 \\
\hline
1
\end{array}
$$

1.

$$8\ \overline{)\ \$64.83} \qquad 6\ \overline{)\ \$52.77} \qquad 8\ \overline{)\ \$65.43} \qquad 6\ \overline{)\ \$49.16}$$

2.

$$9\ \overline{)\ \$11.86} \qquad 5\ \overline{)\ \$38.54} \qquad 4\ \overline{)\ \$35.47} \qquad 7\ \overline{)\ \$64.66}$$

3.

$$8\ \overline{)\ \$89.15} \qquad 2\ \overline{)\ \$50.07} \qquad 6\ \overline{)\ \$94.77} \qquad 8\ \overline{)\ \$97.85}$$

4.

$$6\ \overline{)\ \$33.41} \qquad 2\ \overline{)\ \$12.45} \qquad 7\ \overline{)\ \$42.99} \qquad 9\ \overline{)\ \$62.51}$$

MENU-MATH/ICE CREAM PARLOR

Name _____

Dividing 4-digit numbers by 1 digit, remainder.

1.
2 | $38.25   4 | $26.37   3 | $42.89   8 | $29.37

2.
4 | $21.82   6 | $72.19   5 | $52.78   6 | $25.17

3.
8 | $49.26   7 | $52.93   8 | $52.71   4 | $36.87

4.
7 | $43.13   4 | $11.81   2 | $13.31   5 | $11.92

MENU-MATH/ICE CREAM PARLOR

Name _____

Dividing 4-digit numbers by 1 digit, with and without remainders.

1.

3 | $32.70   4 | $58.70   8 | $45.96   3 | $23.01

2.

2 | $12.48   5 | $34.24   6 | $36.00   7 | $94.66

3.

4 | $19.85   2 | $11.59   7 | $35.63   4 | $99.79

4.

3 | $28.46   5 | $87.55   4 | $82.56   9 | $88.88

MENU-MATH/ICE CREAM PARLOR

Name _____

Dividing 2-digit numbers by 2 digits, no remainder.

EXAMPLE:

$$17 \overline{\smash{\big)}\ \$\ .85} \quad \begin{array}{r} \$\ .05 \\ \hline \\ 85 \end{array}$$

1.

$$12 \overline{\smash{\big)}\ \$\ .48} \qquad 13 \overline{\smash{\big)}\ \$\ .26} \qquad 43 \overline{\smash{\big)}\ \$\ .86} \qquad 31 \overline{\smash{\big)}\ \$\ .93}$$

2.

$$32 \overline{\smash{\big)}\ \$\ .64} \qquad 11 \overline{\smash{\big)}\ \$\ .77} \qquad 23 \overline{\smash{\big)}\ \$\ .69} \qquad 22 \overline{\smash{\big)}\ \$\ .88}$$

3.

$$29 \overline{\smash{\big)}\ \$\ .87} \qquad 14 \overline{\smash{\big)}\ \$\ .42} \qquad 15 \overline{\smash{\big)}\ \$\ .45} \qquad 16 \overline{\smash{\big)}\ \$\ .48}$$

4.

$$17 \overline{\smash{\big)}\ \$\ .68} \qquad 25 \overline{\smash{\big)}\ \$\ .75} \qquad 18 \overline{\smash{\big)}\ \$\ .72} \qquad 19 \overline{\smash{\big)}\ \$\ .57}$$

MENU-MATH/ICE CREAM PARLOR

Name _____

EXAMPLE:

```
        $   .23 R 24¢
    29 | $ 6.91
          5 8
          1 11
            87
            24
```

1.
```
46 | $ 8.02
```
```
16 | $ 4.06
```
```
47 | $ 8.10
```

2.
```
38 | $ 9.10
```
```
49 | $ 9.01
```
```
19 | $ 5.18
```

3.
```
37 | $ 9.08
```
```
39 | $ 9.92
```
```
18 | $ 7.81
```

4.
```
10 | $ 9.75
```
```
22 | $ 6.95
```
```
31 | $ 9.98
```

Name _____

**Dividing 3-digit
numbers by 2 digits,
no remainder.**

_____

1.

$$32 \overline{|\ \$\ 7.68}$$
$$20 \overline{|\ \$\ 8.00}$$
$$14 \overline{|\ \$\ 8.96}$$

2.

$$19 \overline{|\ \$\ 7.98}$$
$$25 \overline{|\ \$\ 5.00}$$
$$16 \overline{|\ \$\ 5.28}$$

3.

$$17 \overline{|\ \$\ 8.50}$$
$$13 \overline{|\ \$\ 1.56}$$
$$16 \overline{|\ \$\ 1.92}$$

4.

$$29 \overline{|\ \$\ 4.35}$$
$$18 \overline{|\ \$\ 3.24}$$
$$12 \overline{|\ \$\ 4.08}$$

MENU-MATH/ICE CREAM PARLOR

Name _____

Dividing 3-digit numbers by 2 digits, remainder.

_____

1.

$19\overline{)\$\,6.21}$ $27\overline{)\$\,6.93}$ $29\overline{)\$\,9.96}$

2.

$38\overline{)\$\,8.92}$ $13\overline{)\$\,8.88}$ $17\overline{)\$\,4.00}$

3.

$75\overline{)\$\,8.64}$ $40\overline{)\$\,8.07}$ $45\overline{)\$\,4.79}$

4.

$82\overline{)\$\,9.09}$ $11\overline{)\$\,2.99}$ $23\overline{)\$\,6.98}$

MENU-MATH/ICE CREAM PARLOR

Name _____

Dividing 4-digit numbers by 2 digits, no remainder.

1.
41 | $10.66   52 | $29.64   43 | $31.39

2.
46 | $17.48   31 | $16.12   32 | $23.36

3.
52 | $40.56   26 | $12.74   16 | $13.28

4.
39 | $31.98   46 | $12.88   27 | $15.12

Name _____

**Dividing 4-digit
numbers by 2 digits,
remainder.**

1.

32 $\overline{)\ \$68.59\ }$ 25 $\overline{)\ \$84.62\ }$ 27 $\overline{)\ \$80.40\ }$

2.

14 $\overline{)\ \$34.28\ }$ 10 $\overline{)\ \$29.82\ }$ 31 $\overline{)\ \$78.72\ }$

3.

36 $\overline{)\ \$82.64\ }$ 15 $\overline{)\ \$63.71\ }$ 42 $\overline{)\ \$97.30\ }$

4.

16 $\overline{)\ \$61.96\ }$ 23 $\overline{)\ \$48.71\ }$ 38 $\overline{)\ \$80.25\ }$

                   MENU-MATH/ICE CREAM PARLOR

Name _____

**Dividing 5-digit
numbers by 2 digits,
no remainder.**

1.

$18\overline{)\ \$\ 900.00}$ $25\overline{)\ \$\ 750.00}$ $12\overline{)\ \$\ 720.00}$

2.

$16\overline{)\ \$\ 800.00}$ $29\overline{)\ \$\ 580.00}$ $32\overline{)\ \$\ 960.00}$

3.

$14\overline{)\ \$\ 700.00}$ $30\overline{)\ \$\ 900.00}$ $10\overline{)\ \$\ 100.00}$

4.

$17\overline{)\ \$\ 680.00}$ $15\overline{)\ \$\ 750.00}$ $26\overline{)\ \$\ 780.00}$

Name _____

Dividing 5-digit numbers by 2 digits, remainder.

1.

59 | $ 181.26   41 | $ 247.16   49 | $ 416.83

2.

71 | $ 573.58   69 | $ 140.94   31 | $ 187.02

3.

61 | $ 341.92   79 | $ 401.23   71 | $ 531.52

4.

21 | $ 113.52   59 | $ 122.89   41 | $ 330.65

MENU-MATH/ICE CREAM PARLOR

Name _____

Multiplying and dividing to solve problems (answers exclude sales tax).

DIRECTIONS:
Solve the following problems. Write an answer in each box. Use your menu.

ORDER	COST	DIVIDED	COST EACH
1. 3 Giant Sundaes		5 Ways	
2. 6 Berry Banana Splits		4 Ways	
3. 8 Golden Cakes		5 Ways	
4. 10 American Banana Splits		5 Ways	
5. 4 Royal Banana Splits		8 Ways	
6. 1 Rainbow Sundae		5 Ways	
7. 4 Apple Banana Splits		5 Ways	
8. 7 Tin Roof Sundaes		5 Ways	
9. 4 Big Scoop Sundaes		2 Ways	
10. 3 Triple Sundaes		4 Ways	
11. 3 Tin Roof Sundaes		5 Ways	
12. 6 Pecan Banana Splits		8 Ways	
13. 6 Raspberry Freezes		3 Ways	
14. 4 Hot Fudge Sundaes		8 Ways	

     MENU-MATH/ICE CREAM PARLOR

Name _____

Multiplying, adding, and dividing to solve problems (answers exclude sales tax).

DIRECTIONS:
Solve the following problems. Write an answer in each box. Use your menu.

ORDER	COST	DIVIDED	COST EACH
1 Triple Sundae 1 Fancy Sundae 2 small Colas Total:	_____ _____ _____ _____	Two Ways	
2 pieces of Fudge Cake 2 pieces of White Cake 5 Hot Teas Total:	_____ _____ _____ _____	Five Ways	
3 Nutty Sundaes 2 Pecan Banana Splits 5 regular Sprites Total:	_____ _____ _____ _____	Five Ways	
4 Orange Freezes 2 Giant Sundaes 1 Root Beer Float Total:	_____ _____ _____ _____	Four Ways	
1 Royal Banana Split 2 Hot Chocolates 2 Coffees 2 Apple Banana Splits Total:	_____ _____ _____ _____ _____	Seven Ways	

Name _____

Solving division story problems.

Directions: Solve the following story problems without the menu.

_____

1. Jerry earned $880.00 at the Old Fashioned Ice Cream Parlor. He put half of what he earned in the bank. How much money did he put in the bank?   _____

2. Four girls went to the Parlor for banana splits. The bill totaled $12.36. Splitting the cost equally, how much should each girl pay?   _____

3. Alan worked at the Old Fashioned Ice Cream Parlor. He made $250.00. He spent half of his earnings for a new 10-speed bike. How much did his bike cost?   _____

4. The cheerleaders had ice cream after the game. Their bill was $13.20. How much should each girl pay if they split the bill six ways?   _____

5. Skip, Ted, and Randy spent $12.30 for ice cream. The boys shared the cost of the check equally. How much did each boy pay?   _____

6. The bill for ice cream comes to $26.55. If nine people share the cost equally, what will each person pay?   _____

7. Mrs. Grant paid $28.35 for 5 gallons of chocolate ice cream. How much did she pay for each gallon?   _____

8. Tony earns $5.00 per hour at the Ice Cream Parlor. How many hours must he work to earn $125.00?   _____

Name _____

Solving division story problems.

Directions: Solve the following story
problems without the menu.

_____

1. Sally worked six days last week. Her paycheck
   for the week amounted to $139.92.
   How much per day did Sally earn?   _____

2. Root beer for 16 team members was $15.20.
   How much did each root beer cost?   _____

3. There was a party at the Old Fashioned Ice
   Cream Parlor for 13 people. The check came to $58.50.
   What was the cost for each guest?   _____

4. Mr. Kirk paid $45.50 for sundaes for 14 Boy Scouts.
   What did he pay for each boy's sundae?   _____

5. Ice cream was $21.34 for 11 quarts.
   What was the cost of one quart?   _____

6. Last year the Parlor used 7,200 gallons of ice
   cream. How much ice cream was used in one month?
   (1 year = 12 months)   _____

7. One day the Parlor paid $33.60 for 96 pounds of
   bananas. How much was the cost of one pound of
   bananas?   _____

8. Sara's dad treated eight Girl Scouts to ice cream.
   The bill was $26.00. How much for each girl?   _____

Name _____

Solving division story problems.

Directions: Solve the following story
problems without the menu.

_____

1. The Old Fashioned Ice Cream Parlor uses cherries
   on banana splits, sundaes, and sodas. Last week the
   Parlor spent $74.50 for 50 jars of cherries.
   How much did one jar of cherries cost?   _____

2. Fifty cans of whipped cream costs the Parlor
   $45.00. What is the price of one can?   _____

3. Ted works at the Parlor. Last week he worked 40 hours
   and made $170.00. How much money did Ted earn
   per hour?   _____

4. Dana worked a total of 98 hours in July. She
   earned $431.20. How much does Dana earn per hour? _____

5. Brandon was paid after working at the Parlor
   for 16 days. His paycheck totaled $388.80.
   How much was Brandon paid per day?   _____

6. Karen earned $3,420.00 working at the Ice
   Cream Parlor last year. How much was that
   a month? (1 year = 12 months)   _____

7. The Parlor paid $99.00 for 18 large cans of
   chocolate syrup. How much did each can cost?   _____

8. The manager spent $144.00 for 12 jars of
   nuts. How much did each jar cost?   _____

Name _____

Dividing to find amount of money earned per hour.

Directions:
Figure out how much each employee earns an hour (hourly wage) working at the Ice Cream Parlor.

Example:

```
              $ 4.25  hourly wage
hours worked 34 │ $144.50  amount earned
              136
              ─────
              8 5
              6 8
              ─────
              1 70
              1 70
              ─────
```

EMPLOYEES	HOURS WORKED	HOURLY WAGE	AMOUNT EARNED
1. Sherri	28	_____	$133.00
2. Chuck	29	_____	$123.25
3. Fred	38	_____	$163.40
4. Steve	36	_____	$156.60
5. Connie	24	_____	$111.60
6. Sandra	32	_____	$ 99.20
7. Tom	37	_____	$162.80
8. Dean	40	_____	$171.20
9. Andy	22	_____	$ 95.70
10. Jackie	35	_____	$148.75
11. Tim	34	_____	$144.50
12. Don	26	_____	$111.80

Name _____

Adding, subtracting, multiplying, and dividing to solve story problems (answers exclude sales tax).

Directions: Solve the following story problems using the menu.

1. You and a friend order 2 pieces of White Cake and 2 cups of Hot Chocolate.

    a) What will the check be?   _____

    b) How much change should you get from $10.00?   _____

    c) How much will each of you pay if you split the bill two ways?   _____

2. There are four people at the Old Fashioned Ice Cream Parlor. They order 4 Royal Banana Splits, 2 small Colas, and 2 small Root Beers.

    a) What will be the total of their check?   _____

    b) How much change should they get from $20.00?   _____

    c) How much will each person pay if they split the bill four ways?   _____

3. Dan, Dave, and Doug took their dates to the Old Fashioned Ice Cream Parlor. They ordered 3 pieces of Fudge Cake, 3 Vanilla Super Shakes, one Root Beer Float, and two Raspberry Freezes.

    a) How much will they owe altogether?   _____

    b) Splitting the bill three ways, how much will each person pay?   _____

    c) Dan uses $10.00 to pay his share, how much change will he get back?   _____

4. Five people order the following: 2 Fancy Sundaes, 2 Lime Freezes, 1 Special Sundae, 5 Hot Teas.

    a) What will the check be?   _____

    b) If they divide the check total equally, how much will each person owe?   _____

Name _____

Adding, subtracting, multiplying, and dividing to solve story problems (answers exclude sales tax).

Directions: Solve the following story problems using the menu.

1. Members of the Ski Club stop for cake and hot chocolate. They order the following: 4 pieces of Fudge Cake, 3 pieces of Golden Cake, 1 piece of White Cake, and 8 cups of Hot Chocolate.

    a) What will be the check total? _____

    b) What change should they get from $30.00? _____

    c) If they split the bill eight ways, how much should each member pay? _____

2. Mrs. Scott, Mrs. Grable, and Mrs. Perkins took their children for milk shakes. They ordered the following: 3 Cherry Shakes, 4 Chocolate Super Shakes, and 2 Vanilla Super Shakes with malt.

    a) What will the check be? _____

    b) If the check is paid with $40.00, what change will be received? _____

    c) The mothers will divide the bill equally. How much will each mother owe? _____

3. You pick up an order at the Old Fashioned Ice Cream Parlor for a party. The order includes: 8 Nutty Sundaes, 3 Special Sundaes, 4 Hot Fudge Sundaes, and 15 small Colas.

    a) How much will this order cost? _____

    b) You give the cashier five $10.00 bills and two $5.00 bills. How much should you get back in change? _____

    c) The 15 people attending the party will divide the bill equally. How much will each person owe? _____

Name _____

1. Use the menu to find prices.
2. Multiply to find item costs.
3. Add check to find subtotal.

4. Use tax table to find amount of tax.
5. Add the subtotal and tax.
6. Divide total to find cost per person.

Ice Cream Parlor		
2	Berry Banana Splits	
2	Apple Banana Splits	
3	Vanilla Super Shakes	
	Subtotal	
	Tax	
	Total	

3 people share the cost.

How much does each person pay? _____

Ice Cream Parlor		
7	Hot Fudge Sundaes	
1	Chocolate Super Soda	
	Subtotal	
	Tax	
	Total	

7 people share the cost.

How much does each person pay? _____

Ice Cream Parlor		
3	Root Beer Floats	
1	Rainbow Sundae	
	Subtotal	
	Tax	
	Total	

2 people share the cost.

How much does each person pay? _____

Ice Cream Parlor		
1	Royal Banana Split	
3	Giant Sundaes	
	Subtotal	
	Tax	
	Total	

4 people share the cost.

How much does each person pay? _____

Name _____

1. Use the menu to find prices.
2. Multiply to find item costs.
3. Add check to find subtotal.

4. Use tax table to find amount of tax.
5. Add the subtotal and tax.
6. Divide total to find cost per person.

Ice Cream Parlor		
5	Berry Banana Splits	
3	Coffees	
1	Soda Water	
Subtotal		
Tax		
Total		

7 people share the cost.

How much does each person pay? _____

Ice Cream Parlor		
1	Triple Sundae	
4	Big Scoop Sundaes	
4	Milks	
Subtotal		
Tax		
Total		

5 people share the cost.

How much does each person pay? _____

Ice Cream Parlor		
2	Triple Sundaes	
2	Big Scoop Sundaes	
4	small Colas	
Subtotal		
Tax		
Total		

4 people share the cost.

How much does each person pay? _____

Ice Cream Parlor		
6	Simple Sundaes	
5	regular Root Beers	
1	regular Sprite	
Subtotal		
Tax		
Total		

7 people share the cost.

How much does each person pay? _____

Name _____

1. Use the menu to find prices.
2. Multiply to find item costs.
3. Add check to find subtotal.

4. Use tax table to find amount of tax.
5. Add the subtotal and tax.
6. Divide total to find cost per person.

Ice Cream Parlor		
3	Apple Banana Splits	
3	Scotch Sundaes	
5	regular Sprites	
1	Soda Water	
	Subtotal	
	Tax	
	Total	

8 people share the cost.

How much does each person pay? _____

Ice Cream Parlor		
3	Vanilla Super Sodas	
4	Hot Fudge Sundaes	
2	American Banana Splits	
	Subtotal	
	Tax	
	Total	

6 people share the cost.

How much does each person pay? _____

Ice Cream Parlor		
1	Berry Banana Split	
3	Fudge Cakes	
1	Golden Cake	
6	small Colas	
	Subtotal	
	Tax	
	Total	

5 people share the cost.

How much does each person pay? _____

Ice Cream Parlor		
2	Giant Sundaes	
3	Chocolate Super Sodas	
4	Soda Waters	
2	Triple Sundaes	
4	Iced Teas	
	Subtotal	
	Tax	
	Total	

4 people share the cost.

How much does each person pay? _____

Name _____

1. Use the menu to find prices.
2. Multiply to find item costs.
3. Add check to find subtotal.

4. Use tax table to find amount of tax.
5. Add the subtotal and tax.
6. Divide total to find cost per person.

Ice Cream Parlor

3	Royal Banana Splits	
2	Cola Super Floats	
2	Milks	
1	Soda Water	
	Subtotal	
	Tax	
	Total	

5 people share the cost.
How much does each person pay? _____

Ice Cream Parlor

3	Big Scoop Sundaes	
4	Triple Sundaes	
2	Nutty Sundaes	
	Subtotal	
	Tax	
	Total	

5 people share the cost.
How much does each person pay? _____

Ice Cream Parlor

3	Grand Banana Splits	
4	Special Sundaes	
3	Orange Freezes	
	Subtotal	
	Tax	
	Total	

3 people share the cost.
How much does each person pay? _____

Ice Cream Parlor

2	Fancy Sundaes	
1	Giant Sundae	
1	Scotch Sundae	
2	Apple Banana Splits	
	Subtotal	
	Tax	
	Total	

4 people share the cost.
How much does each person pay? _____

Name _____

1. Use the menu to find prices.
2. Multiply to find item costs.
3. Add check to find subtotal.

4. Use tax table to find amount of tax.
5. Add the subtotal and tax.
6. Divide total to find cost per person.

Ice Cream Parlor		
4	small Root Beers	
2	Royal Banana Splits	
2	Vanilla Super Shakes	
2	Cola Super Floats	
	Subtotal	
	Tax	
	Total	

6 people share the cost.

How much does each person pay? _____

Ice Cream Parlor		
3	American Banana Splits	
5	regular Colas	
3	Coffees	
	Subtotal	
	Tax	
	Total	

4 people share the cost.

How much does each person pay? _____

Ice Cream Parlor		
5	White Cakes	
3	small Root Beers	
2	Lime Freezes	
	Subtotal	
	Tax	
	Total	

5 people share the cost.

How much does each person pay? _____

Ice Cream Parlor		
3	Vanilla Super Shakes	
3	Cherry Super Shakes	
3	regular Colas	
	Subtotal	
	Tax	
	Total	

2 people share the cost.

How much does each person pay? _____

Name _____

1. Use the menu to find prices.
2. Multiply to find item costs.
3. Add check to find subtotal.

4. Use tax table to find amount of tax.
5. Add the subtotal and tax.
6. Divide total to find cost per person.

Ice Cream Parlor		
1	Giant Sundae	
2	American Banana Splits	
1	Vanilla Soda	
3	Milks	
Subtotal		
Tax		
Total		

5 people share the cost.

How much does each person pay? _____

Ice Cream Parlor		
2	Simple Sundaes	
3	Pineapple Sodas	
2	Root Beer Floats	
1	Cherry Shake	
1	Fudge Cake	
Subtotal		
Tax		
Total		

8 people share the cost.

How much does each person pay? _____

Ice Cream Parlor		
1	Nutty Sundae	
3	Raspberry Freezes	
2	Fancy Sundaes	
1	American Banana Split	
2	Royal Banana Splits	
Subtotal		
Tax		
Total		

4 people share the cost.

How much does each person pay? _____

Ice Cream Parlor		
1	Scotch Sundae	
2	Lime Freezes	
2	Pineapple Super Sodas	
1	Strawberry Shake	
1	Chocolate Milk	
Subtotal		
Tax		
Total		

7 people share the cost.

How much does each person pay? _____

Name _____

POST TEST — MULTIPLICATION & DIVISION
(answers exclude sales tax)

1.
$$\begin{array}{r} \$\ .85 \\ \times\ 6 \\ \hline \end{array} \qquad \begin{array}{r} \$\ .59 \\ \times\ 9 \\ \hline \end{array} \qquad \begin{array}{r} \$\ 6.26 \\ \times\ 5 \\ \hline \end{array} \qquad \begin{array}{r} \$\ 9.68 \\ \times\ 8 \\ \hline \end{array} \qquad \begin{array}{r} \$\ 13.57 \\ \times\ 7 \\ \hline \end{array}$$

2.
$$\begin{array}{r} \$\ 37.25 \\ \times\ 4 \\ \hline \end{array} \qquad \begin{array}{r} \$\ .39 \\ \times\ 62 \\ \hline \end{array} \qquad \begin{array}{r} \$\ 6.37 \\ \times\ 15 \\ \hline \end{array} \qquad \begin{array}{r} \$\ 8.96 \\ \times\ 82 \\ \hline \end{array}$$

3.
$$\begin{array}{r} \$\ 7.56 \\ \times\ 63 \\ \hline \end{array} \qquad \begin{array}{r} \$\ 9.08 \\ \times\ 70 \\ \hline \end{array} \qquad \begin{array}{r} \$\ 86.58 \\ \times\ 39 \\ \hline \end{array} \qquad \begin{array}{r} \$\ 50.40 \\ \times\ 56 \\ \hline \end{array}$$

4. 32 Apple Banana Splits   @ $ 2.80 ea.   _____

5. Maria makes $4.40 an hour working at the Old Fashioned
    Ice Cream Parlor. She worked 35 hours during one week.
    How much money did she earn for the week?   _____

6. Carol had $20.75 to spend at the Old Fashioned
    Ice Cream Parlor. She bought the following items for
    herself and her friends:

                Three Grand Banana Splits   @ $4.95 ea. _____

                Three regular Sprites   @ $ .95 ea.   _____

    a) How much did she spend?   _____

    b) How much change will she receive?   _____

Name _____

POST TEST — MULTIPLICATION & DIVISION
(answers exclude sales tax)

1.
6 $\overline{)\ \$\quad.96}$ 7 $\overline{)\ \$\quad 8.47}$ 7 $\overline{)\ \$\ 42.00}$

2.
4 $\overline{)\ \$\ 82.52}$ 17 $\overline{)\ \$\quad.85}$ 32 $\overline{)\ \$\quad 2.56}$

3.
26 $\overline{)\ \$\ 12.74}$ 38 $\overline{)\ \$\ 78.51}$ 72 $\overline{)\ \$251.63}$

4. You and a friend order the following: Two Fudge
    Cakes @ $2.80 each and 2 small Sprites @ $.65 each.

    a) What will the check be?   _____

    b) How much change should you get from $10.00?   _____

    c) How much will each of you pay if the bill is
        divided two ways?   _____

5. The basketball team had ice cream after the game.
    Their bill was $24.24. How much should each player pay
    if they split the bill six ways?   _____

Ice Cream Parlor — Answer Key

PAGE 1:					
1)	.44	.78	.66	1.05	
2)	.84	.64	.36	.52	
3)	.50	1.05	.56	.80	
4)	1.48	.87	1.60	.72	
5)	1.08	1.00	.82	1.72	
6)	.90	.32	.84	.63	

PAGE 2:					
1)	1.50	1.44	2.48	1.83	
2)	.72	1.84	3.00	2.25	
3)	1.48	1.32	2.55	1.84	
4)	2.43	2.08	3.44	2.20	
5)	3.76	1.35	1.66	4.05	
6)	1.86	1.30	2.12	3.05	

PAGE 3:					
1)	1.32	1.61	1.08	1.68	
2)	1.50	2.17	2.16	2.59	
3)	1.68	.90	2.52	2.73	
4)	1.26	1.26	2.10	3.84	
5)	4.34	3.12	3.18	3.78	
6)	2.88	3.15	3.29	2.87	

PAGE 4:					
1)	2.88	3.43	6.44	5.58	
2)	3.66	4.38	4.02	5.74	
3)	4.48	4.34	6.23	5.28	
4)	4.92	5.88	6.72	5.94	
5)	5.25	5.70	5.11	3.96	
6)	6.16	4.86	3.84	3.99	

PAGE 5:					
1)	1.68	2.07	3.36	4.41	
2)	2.00	2.34	1.20	2.52	
3)	3.15	2.48	3.04	3.69	
4)	2.88	2.16	3.76	3.87	
5)	1.44	1.12	4.16	4.50	
6)	3.24	3.44	5.04	3.06	

PAGE 6:					
1)	5.94	4.40	5.20	5.49	
2)	4.14	3.84	4.56	2.16	
3)	2.61	6.84	5.52	4.24	
4)	4.32	4.48	5.76	5.85	
5)	6.80	3.69	6.08	7.60	
6)	4.72	7.36	8.64	7.56	

PAGE 7:					
1)	12.33	8.68	12.30	11.61	
2)	24.42	40.25	11.00	10.86	
3)	46.05	24.12	14.48	30.80	
4)	34.30	12.04	10.60	12.57	
5)	8.36	6.45	24.72	15.20	
6)	18.06	36.64	30.60	6.27	

PAGE 8:					
1)	15.05	17.88	52.22	16.02	
2)	16.94	20.58	57.96	48.72	
3)	17.92	21.48	52.92	19.38	
4)	32.83	30.18	32.41	51.36	
5)	33.25	41.88	54.90	64.96	
6)	42.70	49.84	63.49	59.58	

PAGE 9:					
1)	19.84	20.70	53.76	64.62	
2)	14.58	23.84	65.07	60.48	
3)	30.78	28.80	73.08	68.00	
4)	37.35	34.24	78.75	72.96	
5)	42.57	46.17	82.62	76.00	
6)	47.04	55.08	77.04	89.28	

PAGE 10:					
1)	26.10	57.16	38.46	45.30	
2)	61.50	42.00	90.08	133.60	
3)	129.80	55.17	37.12	66.20	
4)	94.08	35.85	36.87	91.40	
5)	64.52	70.32	68.58	128.30	
6)	49.66	102.44	55.47	80.91	

PAGE 11:					
1)	75.12	87.71	178.57	313.08	
2)	67.50	83.65	126.91	117.78	
3)	173.95	249.69	194.64	165.72	
4)	141.48	265.23	139.08	179.41	
5)	271.38	295.75	375.48	390.84	
6)	439.50	568.75	556.02	438.13	

PAGE 12:					
1)	290.07	284.96	641.34	552.96	
2)	259.44	371.25	661.05	522.00	
3)	269.12	314.28	651.24	514.96	
4)	289.84	521.19	589.44	556.65	
5)	369.68	427.05	659.12	842.04	
6)	210.72	170.64	691.52	866.16	

PAGE 13:					
1)	5.06	3.12	4.62	5.67	
2)	5.72	9.03	7.14	8.50	
3)	8.28	9.76	2.88	5.51	
4)	6.76	5.16	7.44	6.89	
5)	2.99	2.88	8.06	5.94	

PAGE 14:					
1)	20.93	11.68	58.74	25.16	
2)	45.12	25.08	69.72	17.36	
3)	60.06	31.98	12.88	15.12	
4)	13.28	12.74	40.56	58.48	
5)	33.54	17.48	16.12	23.36	

PAGE 15:					
1)	69.16	10.83	20.37	18.26	
2)	41.82	34.44	46.15	22.08	
3)	38.35	46.20	66.36	56.07	
4)	31.96	20.01	32.40	51.66	
5)	19.74	38.54	39.06	11.16	

PAGE 16:					
1)	98.88	50.76	48.76	59.29	
2)	31.68	35.37	68.97	68.00	
3)	39.36	79.05	84.24	81.84	
4)	65.94	71.36	79.05	66.56	
5)	40.04	27.54	71.74	81.51	

PAGE 17:					
1)	767.25	306.24	861.18	371.07	
2)	155.52	720.51	435.69	234.33	
3)	122.67	388.74	309.96	410.88	
4)	564.48	303.45	226.59	263.13	
5)	348.44	312.84	156.96	456.84	

PAGE 18:					
1)	612.36	325.28	249.48	391.96	
2)	333.75	280.44	121.89	200.34	
3)	185.28	400.82	498.68	248.54	
4)	696.18	374.92	103.60	777.71	
5)	157.64	704.25	542.21	460.11	

PAGE 19:					
1)	342.76	190.08	206.72	339.36	
2)	588.98	348.30	273.15	509.76	
3)	161.65	291.60	141.40	564.00	
4)	556.83	231.42	206.64	458.90	
5)	516.60	423.60	304.50	137.19	

PAGE 20:			
1)	2085.50	2590.38	
	1877.12	3837.75	
2)	1992.72	2866.05	
	2305.22	1797.66	
3)	3385.78	3240.37	
	6918.24	1531.92	
4)	2388.88	1413.76	
	5324.56	6679.26	
5)	1445.22	5645.70	
	2203.52	5559.04	

PAGE 21:			
1)	3398.91	3064.95	
	1415.04	4320.60	
2)	3327.28	1967.94	
	4904.55	1740.55	
3)	6286.32	2504.16	
	3868.55	5639.22	
4)	3975.00	6942.00	
	5624.00	4002.00	
5)	3596.00	1862.00	
	1872.00	5443.20	

PAGE 22:			
1)	1243.10	513.28	
	7283.64	1241.86	
2)	2941.26	900.60	
	541.08	1616.16	
3)	4920.82	1349.12	
	1251.25	1514.52	
4)	1805.40	3537.60	
	1379.40	2325.80	
5)	723.60	851.36	
	1142.40	4737.70	

PAGE 25:		
1)	1.02	15.60
2)	1.12	17.05
3)	1.17	17.92
4)	1.29	19.78
5)	1.39	21.31
6)	1.55	23.73
7)	1.68	25.65
8)	1.70	25.95
9)	1.73	26.38
10)	1.75	26.78
11)	1.82	27.83
12)	1.95	29.85
13)	1.98	30.31
14)	2.08	31.75
15)	2.12	32.47
16)	2.17	33.19

PAGE 26:			
1)	9.55	2)	18.95
3)	72.90	4)	16.06

PAGE 27:			
1)	16.00	8)	6.00
2)	11.40	9)	65.00
3)	34.65	10)	71.05
4)	62.55	11)	33.60
5)	57.20	12)	43.80
6)	50.40	13)	24.75
7)	155.40	14)	27.00

PAGE 28:		
1)	2.80	16.80
2)	3.05	30.50
3)	2.60	39.00
4)	3.15	69.30
5)	2.80	70.00
6)	2.10	63.00
7)	12.95	77.70
8)	2.75	22.00
9)	2.25	94.50
10)	3.65	182.50
11)	2.25	74.25
12)	2.80	44.80
13)	3.15	56.70
14)	2.60	96.20

PAGE 29: 1) 34.75 2) .35 3) 25.20 4) 5.85
5) 90.65 6) 11.20 7) 10.80
8) 23.25 9) 29.45

Ice Cream Parlor — Answer Key

PAGE 30: 1) 36.85 2) 18.00 3) 25.55 4) 20.10 5) 14.40 6) 40.15 7) 53.00

PAGE 31: 1) 1.10 2) 12.00, 8.00 3) 25.80, 24.20 4) 13.40, 6.60 5) 15.80, 4.20 6) 40.15, 6.35

PAGE 32: 1) 78.00 2) 136.00 3) 764.50 4) 132.50 5) 157.50 6) 465 7) 400 8) 185.60 9) 384

PAGE 33:
1) 98.90   7) 150.15
2) 110.00   8) 137.95
3) 76.50   9) 156.60
4) 146.20   10) 161.00
5) 121.50   11) 72.25
6) 165.75   12) 111.80

PAGE 34:
1) 6.40 8.40 5.70 20.50
2) 13.00 8.40 7.20 28.60
3) 7.30 8.40 3.75 19.45
4) 11.20 8.10 4.90 24.20
5) 14.60 9.00 5.20 28.80
6) 13.50 4.50 5.20 23.20
7) 15.75 13.90 5.85 35.50
8) 12.00 4.20 3.25 19.45

NOTE: For restaurant checks, answers shown include: total amount/change due

PAGE 35:
15.57/4.43   18.46/1.64
21.03/3.97   21.35/8.65

PAGE 36:
21.61/3.39   20.06/.94
22.74/7.26   21.67/3.33

PAGE 37:
31.08/8.92   28.36/21.64
23.91/6.09   19.58/.42

PAGE 38:
31.46/3.54   17.66/2.34
23.11/1.89   25.20/4.80

PAGE 39:
33.81/6.19   22.68/7.32
32.74/17.26   22.04/2.96

PAGE 40:
34.19/5.81   34.03/.97
31.46/8.54   33.81/16.19

PAGE 41: 1) 3.80, 3.24, 3.42, 24.76, 64.33 2) 50.00, 54.12, 436.17, 436.68 3) 26.04, 64.78, 68.77, 1123.84, 1500.00 4) 19.20 5) 64.40 6) 388.50 7) 14.85 8) 470.40, 479.60

PAGE 42:
1) .21 .12 .16 .24
2) .16 .15 .29 .12
3) .18 .14 .48 .17
4) .16 .10 .11 .21
5) .14 .13 .25 .32

PAGE 43:
1) .18 .12 .42 .17
2) .19 .21 .15 .30
3) .32 .32 .10 .11
4) .13 .32 .11 .13
5) .15 .12 .16 .11

PAGE 44:
1) .30 R2 .24 R1 .12 R1 .11 R4
2) .11 R7 .10 R2 .12 R2 .43 R1
3) .11 R3 .13 R5 .10 R8 .16 R2
4) .28 R1 .26 R2 .22 R1 .11 R2
5) .20 R1 .35 R1 .29 R1 .13 R3

PAGE 45:
1) 4.05 1.00 3.24 1.61
2) 2.08 1.60 2.71 1.11
3) 1.73 1.19 1.21 1.43
4) 1.21 1.20 1.31 1.51

PAGE 46:
1) .47 R4 2.10 R3 .65 R4 1.13 R3
2) .65 R4 2.32 R2 .20 R5 1.41 R3
3) .98 R5 2.88 R1 2.07 R2 1.29 R6
4) .93 R4 .89 R3 3.63 R1 .56 R7

PAGE 47:
1) .65 .50 R3 .89 .85 R2
2) .22 .80 R2 .75 3.00 R1
3) .30 1.05 R3 1.60 1.10 R5
4) 1.12 2.43 1.76 R1 4.26 R1

PAGE 48:
1) 33.00 11.00 22.00 22.00
2) 10.00 44.00 21.00 12.00
3) 3.00 11.00   6.00   8.00
4) 5.00   7.00 23.00   8.00

PAGE 49:
1) 17.71 19.72 13.81 37.84
2) 13.71 19.75   6.13   8.60
3) 5.58   7.49   9.74   8.70
4) 6.95   6.24   5.09 20.64

PAGE 50:
1) 8.10 R3 8.79 R3 8.17 R7 8.19 R2
2) 1.31 R7 7.70 R4 8.86 R3 9.23 R5
3) 11.14 R3 25.03 R1 15.79 R3 12.23 R1
4) 5.56 R5 6.22 R1 6.14 R1 6.94 R5

PAGE 51:
1) 19.12 R1 6.59 R1 14.29 R2 3.67 R1
2) 5.45 R2 12.03 R1 10.55 R3 4.19 R3
3) 6.15 R6 7.56 R1 6.58 R7 9.21 R3
4) 6.16 R1 2.95 R1 6.65 R1 2.38 R2

PAGE 52:
1) 10.90 14.67 R2 5.74 R4 7.67
2) 6.24 6.84 R4 6.00 13.52 R2
3) 4.96 R1 5.79 R1 5.09 24.94 R3
4) 9.48 R2 17.51 20.64 9.87 R5

PAGE 53:
1) .04 .02 .02 .03
2) .02 .07 .03 .04
3) .03 .03 .03 .03
4) .04 .03 .04 .03

PAGE 54:
1) .17 R20 .25 R6 .17 R11
2) .23 R36 .18 R19 .27 R5
3) .24 R20 .25 R17 .43 R7
4) .97 R5 .31 R13 .32 R6

PAGE 55:
1) .24 .40 .64
2) .42 .20 .33
3) .50 .12 .12
4) .15 .18 .34

PAGE 56:
1) .32 R13 .25 R18 .34 R10
2) .23 R18 .68 R4 .23 R9
3) .11 R39 .20 R7 .10 R29
4) .11 R7 .27 R2 .30 R8

PAGE 57:
1) .26 .57 .73
2) .38 .52 .73
3) .78 .49 .83
4) .82 .28 .56

PAGE 58:
1) 2.14 R11 3.38 R12 2.97 R21
2) 2.44 R12 2.98 R2 2.53 R29
3) 2.29 R20 4.24 R11 2.31 R28
4) 3.87 R4 2.11 R18 2.11 R7

PAGE 59:
1) 50.00 30.00 60.00
2) 50.00 20.00 30.00
3) 50.00 30.00 10.00
4) 40.00 50.00 30.00

PAGE 60:
1) 3.07 R13 6.02 R34 8.50 R33
2) 8.07 R61 2.04 R18 6.03 R9
3) 5.60 R32 5.07 R70 7.48 R44
4) 5.40 R12 2.08 R17 8.06 R19

PAGE 61:
1) 20.85 4.17 8) 19.25 3.85
2) 16.80 4.20 9) 12.20 6.10
3) 21.60 4.32 10) 9.60 2.40
4) 31.50 6.30 11) 8.25 1.65
5) 12.00 1.50 12) 16.80 2.10
6) 12.95 2.59 13) 13.50 4.50
7) 11.20 2.24 14) 10.40 1.30

PAGE 62:
3.20, 2.60, 1.30, 7.10, 3.55
5.60, 5.40, 3.50, 14.50, 2.90
10.95, 5.60, 4.75, 21.30, 4.26
9.00, 13.90, 2.10, 25.00, 6.25
3.00, 1.50, 1.80, 5.60, 11.90, 1.70

PAGE 63:
1) 440.00 2) 3.09 3) 125.00 4) 2.20 5) 4.10 6) 2.95 7) 5.67 8) 25

PAGE 64:
1) 23.32 2) .95 3) 4.50 4) 3.25 5) 1.94 6) 600 7) .35 8) 3.25

PAGE 65:
1) 1.49 2) .90 3) 4.25 4) 4.40 5) 24.30 6) 285.00 7) 5.50 8) 12.00

PAGE 66:
1) 4.75 2) 4.25 3) 4.30 4) 4.35 5) 4.65 6) 3.10 7) 4.40 8) 4.28 9) 4.35 10) 4.25 11) 4.25 12) 4.30

PAGE 67:
1) 6.90 3.10 3.45
2) 14.60 5.20 3.65
3) 22.95 7.65 2.35
4) 15.65 3.13

PAGE 68:
1) 28.00 2.00 3.50
2) 22.95 17.05 7.65
3) 56.70 3.30 3.78

NOTE: Answers for restaurant checks include: total amount/cost per person.

PAGE 69:
20.49/6.83   22.26/3.18
20.60/10.30   25.52/6.38

PAGE 70:
17.92/2.56   19.90/3.98
16.16/4.04   31.78/4.54

PAGE 71:
24.24/3.03   26.22/4.37
19.05/3.81   33.28/8.32

PAGE 72:
16.75/3.35   31.30/6.26
33.60/11.20   22.36/5.59

PAGE 73:
20.22/3.37   18.08/4.52
21.35/4.27   20.06/10.03

PAGE 74:
19.15/3.83   25.68/3.21
26.48/6.62   17.01/2.43

PAGE 75:
1) 5.10, 5.31, 31.30, 77.44, 94.99
2) 149.00, 24.18, 95.55, 734.72
3) 476.28, 635.60, 3376.62, 2822.40
4) 89.60 5) 154.00
6) 14.85, 2.85, 17.70, 3.05

PAGE 76:
1) .16, 1.21, 6.00
2) 20.63, .05, .08
3) .49, 2.06 R23, 3.49 R35
4) 6.90, 3.10, 3.45
5) 4.04